IN CRISIS,
ON CRISIS

Also by James Cairns

The Democratic Imagination: Envisioning Popular Power in the 21st Century (with Alan Sears)

A Good Book, in Theory: Making Sense Through Inquiry, third edition (with Alan Sears)

The Myth of the Age of Entitlement: Millennials, Austerity, and Hope

IN CRISIS,
ON CRISIS

ESSAYS IN TROUBLED TIMES

JAMES CAIRNS

WOLSAK
& WYNN

Published by Wolsak and Wynn Publishers
280 James Street North
Hamilton, ON L8R2L3
www.wolsakandwynn.ca

Editor: Noelle Allen | Copy editor: Megan Beadle
Cover and interior design: Jennifer Rawlinson
Author photograph: Layne Beckner Grime
Typeset in Adobe Caslon Pro and Futura PT Cond
Printed by Brant Service Press Ltd., Brantford, Canada

Excerpt from *Moon of the Crusted Snow: A Novel*, Waubgeshig Rice, published by ECW Press Ltd., 2018. 9781770414006. Courtesy of author and ECW Press.

The publisher gratefully acknowledges the support of the Canada Council for the Arts and the Ontario Arts Council. We also acknowledge the financial support of the Government of Canada through the Canada Book Fund and the Government of Ontario through the Ontario Book Publishing Tax Credit and Ontario Creates.

Library and Archives Canada Cataloguing in Publication

Title: In crisis, on crisis : essays in troubled times / James Cairns.
Names: Cairns, James Irvine, 1979- author.
Identifiers: Canadiana 20250178486 | ISBN 9781998408191 (softcover)
Subjects: LCSH: History, Modern—21st century. | LCGFT: Essays.
Classification: LCC PS8605.A4178 I6 2025 | DDC C814/.6—dc23

To Jess, Gus and Winnie

The world keeps ending, and the world goes on.
– Franny Choi

CONTENTS

THIS IS A CRISIS, THIS IS NOT A CRISIS

In October 2020, the poet Sabrina Orah Mark wrote: "It's now one thousand and one o'clock. It's now later than we ever thought."[1] Her focus was on the looming US election, but her foreboding was fuelled by disasters all around. A modern plague, the climate crisis, fascist street violence, the worst global economic collapse since the Great Depression. As I looked out from the 2020 lockdown, I saw images of empty airports, refrigerated trucks filled with the Covid dead, long lines at food banks, another cop murdering another Black man – his name was George Floyd – and righteous flames swallowing a police precinct in the dark.

Even before the novel coronavirus moved from bat to something-or-other to human, the essayist Elisa Gabbert declared that "this point in history does feel different, like we're nearing an event horizon."[2] Way back in 2015, the cultural critic Olivia Laing recalls thinking to herself that "every crisis, every catastrophe, every threat of nuclear war was instantly overridden by the next."[3] Who's lived through the past two decades and not felt history shifting beneath their feet? (At least in the West, where stability is widely taken for granted.) The death of the welfare state, the rise of the far right and mass distrust in political representatives are among the signs of political decay leading many to conclude that democracy itself is in crisis. The political scientist Thea Riofrancos says even this lens is too rosy. Better to understand the twenty-first century as "interlocking crises" of politics, economics and ecology.[4] The *Collins Dictionary* Word of the Year in 2022 was "permacrisis," defined as: "An extended period of instability and insecurity, especially one

1

resulting from a series of catastrophic events."[5]

What do I say to my two young children, to my political allies, to my students – to myself – about how to live through such turmoil, about how the crises of our times end? Are the crises of our times more plentiful, more dangerous, than crises of the past? Have we truly reached a breaking point? How do we orient to the future? What do I do with the sense of powerlessness and fear I share with so many others? The philosopher Tangjia Wang argues that the nature of crisis is to cause simultaneously fear and anxiety, as well as "anticipation and hope."[6] Do I, do we, find hope in our times of crisis?

These essays, written over the past seven years, focus on crisis as a concept, as a calamitous event and as subjective feelings, if not of living through catastrophe, then of standing at the crossroads between ruination and salvation. I take up debates about climate change, war, economic inequality, democratic backsliding and related issues because I'm concerned about humanity's prospects. However, I'm also interested in how crisis is experienced at the individual level, how crisis is felt and talked about (with others and in our own heads), both globally, over years, and in the personalized pinpricks of day-to-day time. At a moment when crisis-talk is everywhere but detailed analysis of crisis remains rare, this book helps think about what we mean by crisis, and what crises mean to us.

Is crisis even the right way to frame contemporary world events? Real shifts are happening, no doubt about that; but do they add up to real crises? Economic crashes aren't unique; on the contrary, they're inherent to capitalism. Voter rates are middling but social movements are flourishing. Democracies around the world face weighty challenges; but "democracy has been sorely challenged throughout its history," and it's debatable whether broad democratic decline, to say nothing of democratic crisis, is occurring.[7] According to the World Health Organization, the Covid-19 global emergency ended on May 5, 2023.[8] Whatever is happening to the environment has been happening for decades. Why are we in crisis now, but we weren't before? When did it start? You might ask the same of the *Lancet*, a leading medical journal, which published an article labelling racism a public health crisis in 2020.[9] But look: the National Association for the Advancement of Colored People (NAACP) named its magazine *The Crisis* in 1910. What good is the concept of crisis if the moment of truth is still generations off or has been in a state of suspended animation for more than one hundred years? This is a crisis, this is not a crisis.

What do we mean by the word "crisis"? In everyday conversations, we use the word to refer to markedly different things: a terminal illness, a spot of trouble, chronic problems in a marriage, shoddy goaltending, the end of the world. The academic literature hardly offers greater precision. There are political scientists who declare a crisis whenever electoral participation rates drop a few points. Then there's Reinhart Koselleck, whose 1959 opus *Critique and Crisis* argues that crisis is the condition of modernity. Koselleck says that during the European Enlightenment of the seventeenth and eighteenth centuries, when the authority of the church collapsed and critical reasoning spread, a new era began in which our break with the past, and our belief in human perfectibility, ensured that we would live in constant crisis. We have the power to act on history, to transform the world so that experience matches potential; yet we're condemned to live in a world knowing that such a resolution hasn't been achieved.

In her book *Anti-Crisis*, the philosopher Janet Roitman argues that journalists and scholars, who see crises everywhere, consistently fail to define what they mean by the term. Crisis is taken to be "the starting point for narration," not a condition requiring explanation.[10] The concept itself "is a blind spot that enables the production of knowledge."[11] In *The Crisis Paradigm*, Andrew Simon Gilbert assumes that there is nothing in real-world situations that makes them crises or not. Differences of ideology and methods of conceptualization are too plentiful and run too deep for crisis to be a shared, stable term for describing events in themselves. Take the genocidal Israeli assault on Occupied Gaza beginning in 2023. The United Nations Secretary-General says Israel created a humanitarian crisis in Palestine. By contrast, the Israeli government says it fought a just war in response to the existential crisis of the Jewish state imposed by external enemies. Gilbert says these kinds of contradictory usages are unavoidable and cannot be resolved by appealing to shared definitions rooted in tests of empirical validity. You cannot determine a crisis by any objective measure. The pivotal question, says Gilbert, is whether the concept has been applied in ways that others accept as accurate and meaningful, leading them to adopt such application. In other words, he says, crises don't exist outside of human representation. The study of crisis, in the Roitman-Gilbert view, is primarily the study of language, persuasion, storytelling.

The power to name an event a crisis, however, to employ what Gilbert calls "the crisis paradigm," is real and has material consequences. In Gilbert's words, calling something a crisis "suggests some sort of decisive and corrective response is urgently required."[12] Political adversaries strive to define the crisis in particular ways, "and thereby validate their agendas and projects, lest their opponents do the same." When poor people migrate to rich countries, political elites say we must fix the border crisis. Build higher walls and larger jails, they say, develop new weapons for keeping the heathens out. By contrast, in the words of Nick Dearden, an advocate for migrant rights based in the UK: "What we call a 'migrant crisis,' is actually a crisis of global injustice caused by war, poverty and inequality."[13] The only ethical solution to the crisis is freedom of movement for everyone.

It's hard to think of a concept other than crisis that so efficiently condenses empirical claim (the uniquely unstable time exists *now*), value judgment (it is bad) and demand for action (we must do such and such). Crying crisis is a weapon in battles of interpretation that frame standards of success and failure, inevitability and openness, heroes, villains and horizons of possibility. Understanding how crisis is used in public debate will strengthen our interventions in struggles defining the shape of the future we want and how to build that future.

If crises are disruptions to business-as-usual, and given that business-as-usual in capitalist democracy means violence, mass suffering and ecological destruction, we might think of crises as periods of hope – at least partially. In Daoism, "crisis, as the possibility of catastrophe, is pregnant with the possibility of fortune."[14] Crises are pathways to transformation. The Greek root of the word "crisis" is the verb *krino/krinein*, meaning to separate, choose, decide, judge. Crisis splits time in two.

When normal breaks down, it becomes easier to see where power truly lies and to experiment collectively with how power might be organized differently. The crisis of war in 1917 led Russians to overthrow Tsarism. The economic crash of 2008–09 was the soil from which the Occupy movement blossomed (and the Tea Party, to be sure). Teens in coastal Louisiana once described to me how their communities developed new networks of solidarity (and collective critical analysis) as a result of living through the dual disasters of Hurricane Katrina and the British Petroleum Deepwater Horizon oil spill.[15] It's not the wreckage of crises that brings me hope but the possibility of rebuilding on the other side.

But do I really *want* a crisis? Where's my humility, my fear? I know a family who fled Syria during the civil war. They lived underground in Lebanon for a year before being allowed into Canada. The youngest of four kids was born in hiding. Her dad's body was broken by Assad's torture squads. This family may never see loved ones back home again. Five million Syrians have been displaced by war since 2010. Millions more refugees from many more places seek safety in the rich countries of Europe. Europe's leaders send boats into the Mediterranean to turn back migrant vessels. Go home, they say, or don't go home, but you can't come here.

The historian Robin D.G. Kelley acknowledges that "the oppressed are rarely in a position to benefit from, or take advantage of the crises that beset their lives."[16] Stop conceiving of crises as opportunities, writes Ståle Holgersen: "they are the enemy."[17] The pandemic, the eco-crisis, the re-emergence of fascism – the crises of our time kill, displace, degrade and starve the world's most vulnerable people. Not only that, but as Holgersen shows, crises are essential to reproducing capitalism. Naomi Klein's *The Shock Doctrine: The Rise of Disaster Capitalism* is a record of political and economic elites using the chaos caused by wars and natural disasters to impose brutal, unpopular, pro-business policy from Iraq to South Asia to New Orleans. Easy for me to cheer on the latest crisis from my post of tenured professor. I'm not at risk of losing my job. I have secure housing, health care, access to food. As I argue in the next essay, "The Worse, the Better," there's something perverse about political theory that finds hope in destruction.

Yet the destruction inherent in crisis does not lead Kelley to embrace the status quo. He thinks that crises have not gone deep enough to force the kind of change we need. "To resist normalization and create a real state of emergency is the task at hand, for it signals a loss of consensus or legitimacy for the dominant group."[18] The problem is not too much crisis, Kelley says, but too little. One thousand and one o'clock could mean the time is right.

Alongside my political concerns, there's a second reason I'm fascinated by crisis – as concept, as lived experience. This piece is personal, and not one I've written about before: that is, my life as an alcoholic. I know that things can suddenly fall apart because, for more than twenty years, I've lived through cycles of sobriety, delusion, collapse, recovery, the new normal of sobriety, relapse and repeat.

I started drinking at age twelve and moved from binge drinking to dependence over the next five or six years. Between the ages of eighteen and twenty-one, I began drinking every day, often alone, always to excess, regularly to the point of throwing up and blacking out. Between twenty-one and twenty-five, I was drunk more often than sober. I lost days and nights to blackouts, hurt myself regularly.

During a single month in summer 2003: I fell ten feet off a bar patio after smashing a pint glass on my friend's head. I fell off my bike, smashing my guitar (which, it would seem, I was attempting to play while cycling) and tearing the skin off one cheek. The other cheek was gashed after I stepped uninvited into a 2:00 a.m. street fight between strangers. I drank three shots of whiskey before my 10:00 a.m. master's thesis defence. The skin on my nose was also one big scab because a week before I dove into a three-foot-deep reflective pool in my thesis supervisor's back garden. I'd been house-sitting, alone, for my supervisor at the time, and I puked on a Persian carpet in his guest room after smashing my face in his water feature.

The month before all this, I'd missed two flights, including one to my mothers' wedding, because of being too drunk to understand time. I lied to everyone I knew, including myself, about my drinking problem.

I started a PhD at age twenty-four. Throughout that fall, I was so sick every morning I'd cry before going to class, promise not to drink for at least one day. And every day I'd be drunk by 4:00 p.m., blacked out by 10:00 p.m. One December day, drunk before noon, walking in snow after writing winter exams, I thought to myself: I will die from drinking. Maybe today, maybe in twenty years. But drinking will kill me. And *still* I didn't think of myself as an alcoholic.

Was this a crisis? If not, what's a crisis? If so, when did it start: When I began drinking every day? When I took my first sip? The minute I was born?

A month after that post-exam walk in the snow, I was on the train from Montreal to Toronto. I'd been having an affair since the week before Christmas. My partner's family was visiting Toronto from the US for the holidays. I claimed to be suffering from depression, therefore needed time alone. I got drunk and snuck around with this other woman, including a few days in Quebec. On the train home, the realization hit me: I'm an alcoholic. It's not just that I like to drink; not just that I'm a bit eccentric; not just that I'm young, impulsive and know how to party. This is an actual thing.

The cliché makes me cringe, but it was a true eureka moment. A thought that had not existed suddenly burst into being. Why then? As soon as it existed, I knew that everything would be different. Not necessarily that I'd stop drinking but that I would never not be an alcoholic again. I was living through a crisis, and I was changing.

If only changing meant no longer being addicted to alcohol. As I discuss in "Blackout," the final essay in this book, I've relapsed and recommitted to sobriety again and again since quitting drinking for the first time twenty years ago. I've learned from my struggle with addiction, but most lessons are either too obvious or too idiosyncratic to write down anywhere but in my diary. In this collection, I'm asking: Are there insights to be found about my crisis of alcoholism in exploring related personal and social crises of truth, ecology, democracy, parenthood and the representation of crisis in literature? Or, rather than start by applying the word "crisis" to experiences I haven't always thought about in those terms, I'm writing to find out what I can learn from approaching formative moments in my life through the lens of crisis.

Anything I know about conditions of sudden collapse, and the potential for radical change, is informed both by my reading of social theory and history, and by my experience moving through periods of out-of-control drinking and day-to-day struggles not to drink. Maybe I'll write a scholarly tract on crisis one day, but I find it difficult even to begin thinking about such a project before working out tensions between the sociological and personal, between the cultural and the autobiographical.

Certainly, there are differences between crisis at the level of society and crisis at the level of the individual. Even the most eclectic, temperamental individual is far more coherent, self-reflective and empowered than even an impossibly coherent, simple society. Individuals possess consciousness; societies don't (even if they possess a kind of self-awareness). Individuals have finite lifespans, and, in the absence of appeals to spirituality, cannot transform into a different entity upon their final rest.

Societies can live for centuries, and the death of one society often begins a new one. Societies are contradictory formations that lack coherent wants, needs, aversions, fears and risks. I don't just mean that people living in society have different tastes. I mean, objectively speaking, a society does not possess a single set of needs and fears. Such things differ depending on where in society we stand. What is the class position, gender, ethnicity, citizenship status and so on from which we experience the world?

Yet I'm hardly the first to integrate self and society in discussing politics. Plato's three-part theory of the soul, which assumes individuals are made up of different proportions of appetite, spirit and logic, corresponds to his model of the ideal society, which is divided into three classes (farmers, soldiers and philosopher kings). The illustration beginning Thomas Hobbes' *Leviathan*, which Hobbes himself helped design in 1651, shows the great leader, the absolute sovereign, towering over the city with a sword in one hand and a staff in the other, but his body is made up of thousands of individual townspeople. Haudenosaunee society is traced back to a woman who fell from the sky.

It's not a coincidence that political theorists talk about the body politic. Whether different groups perform the work of different organs or we're all of us the heart and the brains, individuals are both the basis of society and are shaped by society. What's more, both individuals and society pass through, fall into, endure, crisis. Sometimes this leads to transformation. Sometimes collapse, death. But for whatever reason, and I've implied a few, the idea of crisis is hardly ever explored at the individual and social levels simultaneously.

The geographer and Pulitzer Prize–winning author Jared Diamond hints at such exploration in *Upheaval*, the final book in his trilogy on how civilizations rise and fall. Diamond defines crisis, both in the lives of individuals and of nations, as a significant "moment of truth: a turning point, when conditions before and after that 'moment' are 'much more' different from one another than before and after 'most' other moments."[19] But he abandons conceptual questions there. The book then conducts six case studies of how national crises were resolved.

Academics remarking on the slippery meaning of crisis invariably urge would-be users of the concept to proceed with caution: either define and apply the term rigorously or don't talk about crisis at all. Researchers at the Cascade Institute at Royal Roads University are at the forefront of the terminological debate. In discussion papers, technical reports and op-eds, members of the research team argue for understanding our world as being in a period of *polycrisis*: "A global polycrisis occurs when crises in multiple global systems become casually entangled in ways that significantly degrade humanity's prospects."[20] Overlapping crises, the argument goes, produce greater harm than any single crisis would on its own.

At the opening panel of the 2023 World Economic Forum in Davos, Switzerland, historian Niall Ferguson dismissed the term "polycrisis" because it lacks meaning. "It's just history happening," Ferguson said.[21] The historian Adam Tooze, an early adopter of the term "polycrisis," says those who believe that there's nothing uniquely threatening about the problems of our time need to "beware complacency."[22] There is no single or obvious fix for the polycrisis, and things are on track to get much worse. Then, rejecting Tooze's diagnoses, Gideon Rachman, the *Financial Times* foreign affairs columnist, suggests that polycrisis means nothing "other than – 'there's lots of bad stuff happening simultaneously and one thing can affect another.'"[23]

I'm fascinated by these terminological debates but I'm not out to settle them. What interests me is the wide and varied uses the term "crisis" is put to. Here's the question providing coherence to my eclectic inquiries: What work is done – what is intended, and what is achieved – by talking about crisis in different situations, big and small? To paraphrase two historians of crisis: Who is depicting the situation as a crisis (and who isn't), for what reasons?[24]

The philosophers Will Daddario and Theron Schmidt argue that, in our era of "the *apparently* unresolvable" catastrophes of ecological and human suffering, "we must supply critical thought with creative thought."[25] They call for the rule-breaking, emotional, contradictory space of art to fuel new inquiries from crisis on crisis. Jasper Delbecke, an art theorist, says the essay form is especially well suited for such inquiries. Part science, part philosophy, part art and art criticism, determined to display "the complexity of things *as* complexity" (as opposed to the tidy messaging of political campaigns and peer-reviewed journal articles), the essay was born in, and repeatedly "reveals itself in moments of fundamental existential crisis."[26] To *essay*, writes Delbecke, "to discover new ideas by sharing a process of thinking," is key to how we've worked through crises since Michel de Montaigne puzzled through Renaissance and Reformation in the sixteenth century.[27]

The best essays, says Brian Dillon, "perform a combination of exactitude and evasion."[28] In Dillon's phrase, my inquiry into crisis is a work of *essayism*, written to share a process of thinking about big questions with broad relevance that is self-consciously partial, sometimes oblique, necessarily contradictory and always open to failure, even as I try to get it right. I'm driven by the urge to know what insights might be generated by using the crisis frame expansively, creatively, in unconventional areas.

Several essays start where you might expect: the crisis of democracy; the Covid-19 pandemic; the eco-crisis; the Trump-fuelled crisis of truth. Others are motivated by personal questions (about what it means to be an addict, a parent, a socialist, a lover of literature), and employ the concept of crisis (in times of crisis) as an instrument of life-writing. I do so not only as a form of self-discovery (though, to be sure, I've learned plenty about myself as a result of writing this book) but also to discover what narrating my experiences might tell us about the idea of crisis. Some essays, including this one, I've rewritten numerous times over the past seven years. Others, for example, "The Fall of Cuddling, 2018," a short piece reflecting on my first days of fatherhood, I've left largely untouched since finishing them to convey a glimpse of the world specific to a particular moment now past. Messages in bottles dropped in seas not so very long ago, yet inscribed with the unreproducible insight and naivety of not knowing what we now know. Versions of six essays have appeared in academic and literary journals; five essays are published here for the first time.[29]

What I'm inviting readers to think through in these essays is shaped by what makes me me: most significantly, first, my appeal to left social theory as a guide for my abundant curiosity about, and care for, the world; and second, my experience of trying to be a good father and partner, good teacher and good community member, who struggles with addiction and is haunted by the harms I've caused.

Readers will see things, know things about crisis that lie beyond my grasp. However, the limits of the essay, its rootedness in specific experiences, are among the virtues of the form. To quote Theodor Adorno, "the desire of the essay is not to seek and filter the eternal out of the transitory; it wants, rather, to make the transitory eternal."[30] Montaigne's essays are on the art of conversation, on the greatness of Rome, on cowardice, on thumbs, on conscience, on cannibalism, on sleep, on sadness, on fleeing from pleasures at the cost of one's life and countless other subjects, from gallstones to glory. Yet, Montaigne tells readers in the preface, no matter the scene his words depict, "it is my own self that I am painting."[31] Inasmuch as my essays are also self-portraiture, I've *essayed* personal experiences to provoke in readers fresh views on matters global and local, whether you sympathize with my plight, share my beliefs or find yourself at odds what I've written.

In 2017, a debate at the University of Toronto began with the resolution: "American democracy is in its worst crisis in a generation, and Donald Trump is to blame."[32] Supporting the resolution was then–*New York* magazine provocateur Andrew Sullivan, who argued that "in terms of our liberal democracy and constitutional order, Trump is an extinction-level event."[33] His teammate, E.J. Dionne, a senior fellow at the Brookings Institution, added: "Our country is now as close to crossing the line from democracy to autocracy as it has been in our lifetimes."[34] The pro-side won with 64 percent of audience votes.

Between 2016 and 2020, bestselling books were published

with titles such as *How Democracy Ends*, *How Democracies Die* and *Crises of Democracy*.[35] In 2018, law professor Michelle Alexander wrote in the *New York Times*: "It seems there has been a new crisis roiling our nation nearly every day – a new jaw-dropping allegation of corruption, a new wave of repression at the border, another nod to white nationalism or blatant misogyny, another attack on basic civil rights, freedom of the press or truth itself."[36]

If I may mash up pieces of the liberal, conservative and socialist versions of the Trump-era-as-crisis view, the story goes something like this: The forces leading to Trump's election reflected a democracy already in crisis. What put Trump in power? Racism, sexism, political apathy, showbiz politics and the bankruptcy of the status-quo option, with a little nudge, of course, from the anti-democratic Electoral College. The minute Trump moved into the White House in 2017, he began his assault on democracy in earnest. He banned Muslims from entering the country, stacked courts with conservative judges, rolled back environmental protections, used official state business to line his own pockets. Trump offered a quid pro quo to the leader of Ukraine: our weapons and cash for your attack on my political opponents. Trump lied and cheated. He was impeached late in 2019 but not dispatched. The crisis he created deepened.

By the second half of Trump's first term, debates raged about whether the president was paving the way to fascism, or whether Trump's America was already fascist. Concentration camps appeared on the US-Mexico border. Soldiers attacked peaceful protestors in front of the White House to make way for a presidential photo op. Trump called neo-Nazis who marched with torches through Charlottesville, Virginia, "very fine people," and later told the fascist Proud Boys to "stand by."[37] He mused about a third term (after his self-projected second term) in office at the same time as calling mail-in voting fraudulent. By ignoring health experts and demanding the economy reopen no matter the threat of Covid-19,

Trump ensured that thousands of people would get sick and die unnecessarily. He incited a riot at the Capitol.

Trump's first presidential term was maddening, demoralizing and much else. Was it a crisis of democracy? I'm not so sure, for reasons I explore in depth in the essay "Google Alerts." Recall that the opening days of Trump's wretched first administration also saw the largest street mobilizations for gender justice in US history. The Movement for Black Lives surged to new heights under Trump. Teachers' unions and the Red for Ed campaigns in West Virginia, Los Angeles, Chicago, Nevada and other parts of the country fought austerity and won better work conditions and stronger support for students. The Democratic Socialists of America went from being a scattered grouplet of a few hundred diehards to a national organization boasting seventy thousand members. The broad left hadn't been this large or active since at least the mobilization for global justice in the early 2000s, if not since the 1960s New Left. From socialists campaigning for Bernie Sanders to anarchists defending autonomous zones in Seattle, to communist propaganda in the pages of *Teen Vogue*, new experiments in democracy-from-below thrived in Trump's America.

And even if Americans sometimes seem to forget it, the US is not the world. While Trump promised to lead the fight against women's reproductive rights, millions of women across the globe participated in the International Women's Strike, withdrawing their paid and unpaid labour, halting business as usual. Reproductive rights were won in Ireland in 2018. In 2020, the feminist movement won reproductive rights in the Pope's home country of Argentina. The global movement for climate justice had never been bigger or more militant. Indigenous struggles for land and lives spread across Turtle Island.

Corey Robin, author of *The Reactionary Mind: Conservatism from Edmund Burke to Donald Trump*, says it's a mistake to view Trump's first administration as a break from democracy in

America.[38] The remarkable thing about Trump's first term, Robin argues, is how little he accomplished, how constrained he was by the institutions of American democracy. He never built his great wall, never brought factories back to the Rust Belt, never halted impeachment processes. Trump's rhetoric was uniquely uncouth. But President Richard Nixon got rid of an attorney general, then a deputy attorney general, until finding a lackey in the Department of Justice who would fire the special prosecutor on Nixon's tail. Trump ranted about the witch hunts against him, but failed to sack the special counsel investigating him, Robert Mueller. When the Supreme Court struck down parts of Trump's ban on immigration from Muslim countries, Trump just went along with it. Robin says that in comparison to Nixon's autocratic record, Trump's is "a joke."

The great tyrant Trump was booted by ballots in the presidential election of 2020. Even before the result was formally sanctioned, no one took seriously Trump's delusional claim of victory. No army general or left-wing militia arrested him for attempting to steal the election because everyone could see Trump's hold on power slipping. Notwithstanding Twitter bluster, Trump lacked the social weight to draw his party, or Make America Great Again (MAGA) street fighters, into a real battle to hold onto office. (Don't tell me that a shirtless guy in a Viking hat sitting in the Speaker's chair for a few minutes constitutes a real battle.) He was impeached a second time, and left office in disgrace.

Sleepy Joe Biden, a career politician, a centrist Democrat with a cop for a veep took the reins. And so on and so on the status quo. American democracy saved, if it was ever actually threatened. Where's the crisis?

Yet in November 2020, a few days after even *Fox News* started referring to "President-Elect Joe Biden," *Salon* executive editor Andrew O'Hehir announced that the crisis fuelled by Trumpism persists:

One presidential election, won by assembling an emergency coalition and largely throwing ideology to the wayside, cannot resolve the crisis of democracy. For hundreds of millions of people across the "Western" zone of so-called capitalism and so-called liberal democracy, the economic and political system appears to be a dysfunctional scam that locks in inequality, squelches opportunity and condemns many or most people to precarious lives of wage-slave drudgery, bottomless debt and pointless consumerism. Many of those people are prepared to consider alternatives to the current situation, even when those alternatives appear – at least to the privileged and educated classes – to be warmed-over authoritarian drivel, delivered by a patently insincere second-rate celebrity.[39]

And lo – *this is the voice of my 2020-self, gazing into the crystal ball handed to me by the Ghost of Elections Yet to Come* – despite carrying the taint of electoral defeat, despite being a convicted felon, despite closing in on octogenarianess, Trump is elected president in 2024.

It's terrifying that, unlike 2016, when not even Trump's handlers expected the buffoon to win, this time he takes office with a plan (Project 2025) and Republican majorities in both Houses of Congress. Eager to perform the role of strongman, Trump sat on the stage of his second inauguration and signed fifty executive orders, including ones giving legal force to the Trump campaign's verbal attacks on migrants and transgendered people. Orders like these are frightening not only because they unleash new levels of state violence, and not only because they threaten of new violent laws to come, but because Trump 2.0 emboldens far-right movements.

When the Canadian prime minister dined with Trump in late December 2024, the president-elect mused about annexing Canada. Only weeks later, the US pulled out of the Paris Agreement and the World Health Organization and froze foreign aid to all but Israel and Egypt. If the first weeks of Trump 2.0 fell short of creating actual chaos, they certainly made for the theatre of chaos.

Yet without downplaying real threats embodied by Trump 2.0, it must be said that there are real weaknesses in government-by-MAGA. In the months running up to reinauguration day, Trump's hubristic leadership style cost him one marquee cabinet appointee (so long, Matt Gaetz, we hardly knew you), and a group of senators in Trump's own party handed the president-elect a humiliating loss by defying his orders to suspend the debt ceiling. Open war (over temporary work visas) broke out on social media between the openly-racist wing of MAGA and the Elon Musk–led pro-exploitation-of-migrant-labour tech-bro wing. Corey Robin summarized the trend: "Just as we saw powerlessness and incoherence in Trump's first term, we're seeing it now," even before Trump 2.0's first act of office.[40]

The electoral coalition that led Trump-Vance to victory is comprised of blocs with significantly different material interests.[41] With the unifying fun of disposing of Harris-Walz now over, will the MAGA coalition survive the contradictory pressures of governing? The pro-tariff bloc is sure to be lambasted by the bloc committed to international free markets. Red-hat-wearing Trumpists in the heartland who cheered talk of cutting back government during the election campaign won't be cheering when Musk's Department of Government Efficiency comes for their Social Security cheques, Medicare and Medicaid. And noting these internal fissures says nothing about the obstacles posed by forces hostile to Trump in the administrative state, from the Department of Justice to the Pentagon (or in the streets, from progressive social movements).

Perhaps Trump 2.0 will implement radical policy change with the tenacity of an elite athlete, the adroitness of a great legislator, the peacemaking finesse necessary to unify warring GOP factions. The new president's opening moves surprised even seasoned political observers with their swiftness, plenitude, recklessness and savagery. But before we conclude that the Trump juggernaut is unstoppable, remember: this is Donald Trump we're talking about, right? The guy who "expressed support for hanging his vice president" last time around.[42] The Monday evening (January 27, 2025) White House directive to freeze trillions of dollars in federal grants and loans was rescinded on Wednesday afternoon. Too confusing, vulnerable to legal challenges, ill-conceived, even for President Trump. Contradictory interests and pathologies of mind make it highly unlikely that the MAGA machine will run smoothly for long.

If crisis means a drawn-out state of distress and uncertainty, then okay, the Trump years are a crisis. But that's not what the Trump-is-fascism crowd has always argued. According to that crowd, a 2024 win for Trump would be a turning point: the last free and fair election ever to be held in America. Trump the authoritarian would crush Democrats with force and abolish checks on presidential power. Journalists arrested; the election system demolished. The left would have to go underground. In the words of the *Globe and Mail* columnist Andrew Coyne, Trump's return to power is "a crisis like no other in our lifetimes [. . .] not only for the US but for the world."[43] By the time people try to halt the slide into fascism, says Coyne, they'll find "the democratic levers they might once have pulled to demand change are no longer attached to anything." Is that where the US stands today? Is that where America will be in three years?

The month that Trump moved back into the White House, I saw a bunch of op-eds by the same people who called candidate Trump a fascist now arguing that, with new leaders in place and tweaked campaign slogans, the Democrats can win back control of Congress in the 2026 midterm elections. Congressional Democrats who appear to relish going on MSNBC to call Trump a fascist and claim Musk is carrying out a coup, passively sat through Trump's speech to Congress in March 2025. In we step to the warm bath of politics-as-usual, I guess. Is the US on track to become a failed state? To what extent is it already an oligarchy? These are pertinent questions today; but so were they prior to the rise of Trump and so were they during Joe Biden's presidency.

If the evidence that every aspect of our lives is in an unprecedented crisis is as debatable as I think it is, why do so many people, across the political spectrum, routinely and fervently declare that we are indeed constantly in crisis? It could simply be that Twitter (now Musk's X) hot-take culture has pervaded the public sphere to the point that frustration and anxiety can only be expressed in the strongest available terms. Perhaps people are attracted to the tidy "narrative structure" of crisis, which "reduces a complex world to a binary opposition and a temporal sequence of normalcy, disruption and return to stability."[44] Given that the idea of crisis conjures up images of life and death – things could suddenly get much worse or much better – it may be that some aspect of the human psyche can't help but be drawn to crisis-talk. Elisa Gabbert admits that "when something bad happens, I have a strange instinctual desire for things to get even worse – I think of a terrible outcome and then wish for it. I recognize the pattern, but I don't understand it."[45] Death drive? Power play? An alienated plea to feel something beyond what the market offers and demands of us? What do people get out of making the bold claim that world-historic crisis is upon us?

In her history of capitalist "regimes of accumulation," the New School philosopher Nancy Fraser distinguishes between crises *within* a system and a crisis *of* the system.[46] She says that over the past two hundred years, there have been two crises *within* capitalism so severe that they required "a mutation in the nature of capitalism" to avoid destroying the system as a whole.[47] First, the relatively unregulated "competitive capitalism" of the nineteenth century collapsed in the 1940s, unable to manage the crisis imposed by two World Wars and massive international anti-capitalist movements. Second, the comparatively more regulated post-1945 regime, which Fraser terms "state-managed capitalism," collapsed in the early 1970s amid falling profits and a serious political threat from the New Left.[48]

A crisis *of* capitalism is something different. The crisis of capitalism is the final crisis, the crisis that cannot be resolved through a new regime of accumulation. In the words of the German-Polish revolutionary Rosa Luxemburg, writing in 1915, the crisis *of* capitalism leads either to socialism or barbarism. The internal contradictions of the system become so intense, and the forces driving toward the collapse of the system or toward its transcendence become so overwhelming that the essential features of the system unravel and a whole new social order is established. It's not simply that the rules have changed; we're talking about an entirely different game. Think of the collapse of feudal France and the revolutionary creation of capitalist democracy. Think of the fall of Rome and the centuries of "barbarism" (to use Luxemburg's term) that followed.

The crisis of capitalism does not end in better wages, Medicare for all or 3D printers in every home. It ends in the destruction of capital as a social relation, the end of profit-accumulation as core organizer of our relationship with each other and the earth, the replacement of an entire universe of value, production and access to the things we want and need. I like to imagine it will look a lot like the public library, but in all areas of life, on an international scale: from each according to their love of books (and everything) to each according to their need of books (and everything). But maybe it will look like Cormac McCarthy's *The Road*: the war of each against all fought across a wasteland. Whatever it is, it will be systemic transformation.

The Italian Marxist Antonio Gramsci thought he was witnessing the crisis of capitalism, the final collapse of the whole system, when he wrote from his jail cell in the 1930s that "the crisis consists precisely in the fact that the old is dying and the new cannot be born."[49] In this period of transition, this "interregnum," Gramsci wrote, "a great variety of morbid symptoms appear." Morbid symptoms like Mussolini, mass starvation amid untold wealth, Nazism, and the dehumanization of labour in Henry Ford's factories. Capitalism had developed to the point at which technological and economic productivity far outstripped the laws, forms of governance and cultural values necessary to unleash its productive power. Capitalism was in crisis, and the ground was being prepared for socialism. In the interim, things would, and were, getting weirder and nastier than ever.

Gramsci was wrong. He was not witnessing *the* crisis of capitalism. In Fraser's language, the crisis Gramsci lived through was resolved by "a mutation in the nature of capitalism." Certainly, the high welfare state period profoundly altered the balance of power between capital and labour. Gramsci, who died in 1937, would hardly have recognized features of the state-managed regime of accumulation: legalized unions and robust social safety nets demobilizing the Radical Left across the West. Coal, rail, telecom, health care and other major sectors of the economy taken over by the state and run in the public interest. Even Gramsci's great mind would've had a hard time dreaming up The Beatles, the miniskirt, a TV and fridge in every home. Capitalism was transformed by crisis, but it remained intact.

Then again, might there be something happening below the surface this very moment that, one day, in retrospect, will allow people in the future to see that we're living through a crisis that locates our lives in the same moment as Gramsci's? I remember being shocked as an undergrad to learn that the two World Wars of the twentieth century were not isolated events. *What? Tensions left over from 1918 erupted in 1939?* Even more mind-blowing to my eighteen-year-old self: something about the Franco-Prussian War of 1871 (whatever that was) existed in World War One? Of course, no one alive in 1921 or 1928 or 1937 could have seen how the two wars were part of the same crisis. History was not there yet. World War Two didn't exist. Who are we to say whether Gramsci was right or wrong? We will only know the crisis of capitalism after the fact. Hegel said as much in 1820: "The Owl of Minerva takes its flight only when the shades of night are gathering."[50]

When my first child was around the age of three, he threw an epic tantrum at bedtime. Gus screamed and stomped, pounded fists into pillows. He kicked at the pajamas in my hands. Tears glued his blond hair to red cheeks. I know the parenting books say this is healthy, a normal expression of emotions. But I ache when I see Gus so upset. Part of me is jealous of his lack of inhibitions. Part of me worries about how long the tantrum will last; it is bedtime, after all. Mostly, I just want him to feel better.

Twenty minutes later, the tantrum hadn't eased. Gus's fatigue fuelled his rage. My exhausted partner lifted her hand and gently massaged her breast. Gus's screams stopped almost immediately. He recognized the sign even more quickly than I did: Mom was offering him a boob.

For the first two and a half years of Gus's life, this would've been unremarkable. We would never have reached this level of tantrum because nursing would've de-escalated things long ago. But after a period of Gus nursing less and less, we'd said "goodbye to boobs" more than a month before. The offer of a boob, after weeks of abstinence, confused both Gus and me. Not for one second did I doubt the wisdom of my partner Jess's decision. This is about her body and her unique relationship with Gus. I couldn't understand, then, why the appearance of her breast, and Gus's quiet crawl toward it, made my vision blur. It wasn't just Gus who was quiet: the whole house fell silent. Ringing started inside my head. The heat dropped from my face as the room moved away from me. Panic.

I opened my mouth. No words came out. Gus now lay still against Jess's body. Pulling the door closed, I managed to say, "I . . . go . . ." and felt along the wall for the bathroom. Sitting next to the toilet I began to understand what was happening.

Watching Gus realize what he was being offered, the forbidden boob and the relief it promised, I felt the depth of his surrender. I knew it so well: the moment between realizing that sobriety was over and the first taste of alcohol. Whenever I've returned to drinking after a period of abstaining, in the minutes before I lift a drink, my wrists go floppy, my forehead relaxes. I feel like I might shit my pants. Thoughts spill into consciousness that must've been held back by armies of repressive neurons. "I'm back," I'll say to myself, as I bring the drink to my mouth. I chastise the version of me who kept us (my sober and drunk self) from drinking all this time: "Why would you do this to us?" I drink, hold the alcohol in my mouth, press it between my teeth, imagine it soaking into my tongue, into my sinuses and brain. I sigh and say, "See? This is so much better."

Of course, this isn't what was going on with Gus and Jess. The bond between child and mother, the best way to stop nursing, these are uniquely complicated things. A three-year-old being denied breast milk for a month is not a man struggling with sobriety. But that path leading from the agony of abstinence to the ecstasy of drink, I knew that place exactly. The thrill of being released, the disbelief at how easy it is, the sickening shame of giving in, failing, again. By the time Gus was asleep and Jess left his room, I was in my office, filing papers.

Being an addict who's trying to stay sober makes me think a lot about where the crisis *within* the system goes when its not on the surface, tearing the system asunder. Because even when things are fine – I'm not drinking, not craving alcohol, not at risk of relapsing every time I'm alone – the addiction lives in me. I know this because of how completely it can overwhelm me without warning.

I am in crisis; we are in crisis. We're speeding toward tipping points in the climate system. Wars are growing more frequent and more deadly. Profits are on a yo-yo string, and whatever regime of accumulation resolves the worst economic crash since the Great Depression will immediately, through its ameliorative function, set in motion the dynamics of the next crash. As more and more people turn away from voting, from mainstream parties of liberal democracy, from trust in liberal democracy, anti-democratic forces rise on the right and democracy backslides.

I'm nearly four years beyond the longest I'd previously gone without drinking. I dream of drinking every night; my dream-self tells me sobriety isn't who I am. I am always the drunk, even when I'm not drinking. It's not only that things are fragile: things are collapsing.

We're at the moment of decision.

But every day is the decisive moment. The world won't burst into flames tomorrow or ten years from now, no matter how much carbon dioxide gets pumped into the atmosphere. Blanche Verlie, the environmental scholar and activist, doesn't speak in terms of resolving the climate crisis but of "learning to live with climate change" in view of the fact that "'the' world is not ending, but 'a' world is, and that some worlds need to end in order to allow others room to breathe."[51] We'll face different ecological challenges and opportunities depending on what policies we follow today, but even the worst ones won't terminate the possibility of change toward a more sustainable era. And even the best ones won't vanquish the problem of sustainability. For better or worse, capitalism and liberal democracy are not on the edge of collapse. Variations within these systems, even substantive ones, are not the same as historic transformations.

The fact is, I'm not going to drink today. And if I relapse

tomorrow or ten months from now, I have experience and supports to get me through it. It won't necessarily mean my life is ruined. But it might. There is no curing, no transcending my alcoholism.

This is a crisis. This is not a crisis.

So is it choose-your-own-reality, then? Am I proposing a fresh take on the stale postmodern idea that there is no objective real world? No. I recognize that our access to reality is mediated by language, stories and culture. But, as I argue in "The Real Crisis of Truth," not every truth is as true as every other. Some ways of representing reality are more accurate, more honest, more important than others.

There are crisis tendencies and forces mitigating against crisis. The question is: Where are things shifting drastically in society, in our everyday lives; and where, for better or worse, are things more stable? Because life is full of contradictions, the question of whether we're "in" crisis, where crises begin and end, can only be addressed after defining what vantage point we're looking from, what time-scale and space we're talking about, how general or specific we're claiming to be. This is true of all questions of interpretation. And if this collection of essays demonstrates the importance of clarity in our processes of abstraction, that would be wonderful.

Even more wonderful would be if they help envision the future as an open-ended terrain of struggle. "It is exceptionally difficult to grasp the present as history." Thus begins David McNally's book on the 2008–09 financial crisis. "We tend to think of history as a record of past events, of things that are over and done with. We find it difficult to view our current moment as profoundly historical. Yet, the present is invariably saturated with elements of the future, with possibilities that have not yet come to fruition, and may not do so – as the road to the future is always contested."[52] McNally urges us to think about the present as *a becoming*.

McNally's insight is easily misunderstood as academicized self-help advice. The future is whatever we want it to be! Positive thinking will make your dreams come true! That's not it, though.

Here's how the observation makes sense to me in my struggle with addiction. There is no future in which I drink safely. There just isn't. I wish there were. I tried and failed, spectacularly, dozens of times in the first eight years after acknowledging my drinking problem. For a while, I even had a therapist who urged me not to give up trying to drink moderately. (Or, as I sometimes think of him, my favourite therapist.) I hate the Alcoholics Anonymous (AA) assumption that I am *powerless over alcohol*. I'm saddened to know that I'll never enjoy a safe beer with my kids. But I know from experience that every time I drink, I restart a process that leads directly to serious harm, of myself and others.

No amount of positive thinking, no solemn promise will change this. Does that mean the future is determined? Of course not. I might drink tomorrow, I might not. If I relapse again, I might recover without destroying myself and others, and I might not. Free will is a thing: humans make history, just, as Marx says, not in conditions of their choosing. I can't act on the future. And yet, what I'm doing today will make the future (and will make the present history). If transformative possibilities exist (in my case, not transformation into a safe drinker, but into sustainable sobriety), I must find those possibilities in the world I live in in this moment. The philosopher Ernst Bloch, seeking grounds for hope in socialist revolution, wrote about looking for "tomorrow in today."[53]

Crisis-talk overflows with assumptions about the relationship between now and later, typically by implying that drastic change is imminent. Those who see monumental change around the corner may prove to be prescient. There's a risk, though, that the more rigid our predictions become, the less able we are to see the fundamentally open-ended character of the future. If leftists are wrong about the crisis we say we're in, might we be undercutting our ability to build the very movements necessary to provoke and capitalize on a genuine crisis of the ruling order? Andrew Simon Gilbert reminds us that when elites bemoan the crisis of democracy, the crisis of masculinity, the crisis of culture, the crisis of the nation, they're trying to shore up support for dominant institutions. Crisis-talk can bring events into sharper view, but even when it portrays itself as purely descriptive, it is always a window onto hopes and fears we're trying to realize, avoid or ignore.

Considering threats to all life on Earth, as well as the desires and heartaches that comprise individual lives, I assume that everyone, at one time or other, to some degree, has sensed themselves standing at the decisive turning point. It's unlikely, however, that many of us have had the time or resources necessary for sustained reflection on the character and implications of our crises. These essays are my such reflections. Only after finishing writing them did I begin thinking of the essays as adding up to my provisional, fragmented, meandering answer to a funny-and-terrifying question appearing in the apocalyptic poetry of Franny Choi:

> For years you've kept one eye on the shadows
> swilling about the door, waiting for the arrival
> of the God of Doom. What to do now
>
> that he's here, sipping coffee in our kitchen?[54]

NOTES

1   Sabrina Orah Mark, "It's Time to Pay the Piper," *Paris Review*, October 7, 2020, https://www.theparisreview.org/blog/2020/10/07/its-time-to-pay-the-piper/.

2   Elisa Gabbert, "Vanity Project," in *The Unreality of Memory: And Other Essays* (Farrar, Straus and Giroux, 2020), Kobo.

3   Olivia Laing, "You Look at the Sun," in *Funny Weather: Art in an Emergency* (W.W. Norton, 2020), 1.

4   Thea Riofrancos, "Planet to Win with Thea Riofrancos and Daniel Aldana Cohen," *The Dig*, January 9, 2020, https://thedigradio.com/podcast/planet-to-win-with-thea-riofrancos-and-daniel-aldana-cohen/.

5   Helen Bushby, "Permacrisis Declared Collins Dictionary Word of the Year," *BBC News*, November 1, 2022, https://www.bbc.com/news/entertainment-arts-63458467.

6   Tangjia Wang, "A Philosophical Analysis of the Concept of Crisis," *Frontiers of Philosophy in China* 9, no. 2 (2014): 266.

7   Serge Schmemann, "Democracy Is Not Facing a Global Extinction Event," *New York Times*, January 11, 2025, https://www.nytimes.com/2025/01/11/opinion/editorials/liberal-democracy-far-right-authoritarianism-populism-europe.html.

8   Maria Cheng, "WHO Says COVID Emergency Is Over. So What Does That Mean?," *CTV News*, May 5, 2023, https://www.ctvnews.ca/health/article/who-says-covid-emergency-is-over-so-what-does-that-mean/.

9   Delan Devakumar et al., "Racism, the Public Health Crisis We Can No Longer Ignore," *Lancet* 395, no. 10242 (2020): E112–E113.

10   Janet Roitman, *Anti-Crisis* (Duke University Press, 2014), 34.

11   Roitman, *Anti-Crisis*, 39.

12   Andrew Simon Gilbert, *The Crisis Paradigm: Description and Prescription in Social and Political Theory* (Palgrave Macmillan, 2019), 2.

13   Global Justice Now, "This Is Not a 'Migrant Crisis' – It's a Crisis of Inequality and War," press release, May 9, 2016, https://www.globaljustice.org.uk/news/not-migrant-crisis-its-crisis-inequality-and-war/.

14   Wang, "Philosophical Analysis," 259.

15   James Cairns, "Millennial Blowout: Eco-Disentitlement Versus Ecological Justice," chap. 5 in *The Myth of the Age of Entitlement: Millennials, Austerity, and Hope* (University of Toronto Press, 2017), 109–34.

16   Robin D.G. Kelley, "Crisis: Danger, Opportunity, and the Unknown," *South* 50, no. 1 (2017): 4–5.

17 Ståle Holgersen, *Against the Crisis: Economy and Ecology in a Burning World* (Verso, 2024), 1–20.

18 Kelley, "Crisis," 6.

19 Jared Diamond, "Two Stories," in *Upheaval: Turning Points for Nations in Crisis*, narrated by Henry Strozier (Recorded Books, 2019), audiobook, Libby.

20 Michael Lawrence, Scott Janzwood and Thomas Homer-Dixon, "What Is a Global Polycrisis?: And How Is It Different From a Systemic Risk," Version 2.0, discussion paper 2022-4, Cascade Institute, September 2022, 2, https://cascadeinstitute.org/wp-content/uploads/2022/04/What-is-a-global-polycrisis-v2.pdf.

21 Yasmeen Serhan, "Why 'Polycrisis' Was the Buzzword of Day 1 in Davos," *Time*, January 17, 2023, https://time.com/6247799/polycrisis-in-davos-wef-2023/.

22 Adam Tooze, "Welcome to the World of the Polycrisis," *Financial Times*, October 28, 2022, https://www.ft.com/content/498398e7-11b1-494b-9cd3-6d669dc3de33.

23 Gideon Rachman (@gideonrachman), "'Polycrisis' – now adopted by Davros – is rapidly becoming one of my least favourite cliches. Does it actually mean anything? Other than – 'there's lots of bad stuff happening simultaneously and one thing can affect another,'" Twitter (now X), January 16, 2023, https://twitter.com/gideonrachman/status/1614893315002077184.

24 Rüdiger Graf and Konrad H. Jarausch, "'Crisis' in Contemporary History and Historiography," in Docupedia-Zeitgeschichte, March 27, 2017, 6, https://zeitgeschichte-digital.de/doks/frontdoor/deliver/index/docId/789/file/docupedia_graf_jarausch_crisis_v1_en_2017.pdf.

25 Will Daddario and Theron Schmidt, "Introduction: Crisis and the Im/Possibility of Thought," *Performance Philosophy* 4, no. 1 (2018): 1, 3, italics in original.

26 Jasper Delbecke, "The Essay in Times of Crisis," *Performance Philosophy* 4, no. 1 (2018): 113, 108, italics in original.

27 Delbecke, "Essay in Times of Crisis," 121.

28 Brian Dillon, "On Essays and Essayists," in *Essayism: On Form, Feeling, and Nonfiction* (New York Review of Books, 2017), Kobo.

29 See my "On the Pleasures of Reading the Apocalypse," *Hamilton Review of Books*, November 8, 2022, https://hamiltonreviewofbooks.com/blog/2022/11/08/on-the-pleasures-of-reading-the-apocalypse; "My Struggle and *My Struggle*," *Canadian Notes & Queries* 112 (Fall–Winter 2022–2023): 22–23; "On Fearing Uncertainty: Parenting Amid the Eco-Crisis," *Montreal Review*, July 2023, https://www.themontrealreview.com/Articles/On_Fearing_Uncertainty

_Parenting_Amid_the_Eco_Crisis.php; "Blackout Autotheory," *TOPIA: Canadian Journal of Cultural Studies* 47 (September 2023): 323–44; "The Crisis of Democracy Archive," *Canadian Journal of Communication* 49, no. 1 (2024): 146–58; "The Worse, the Better: A Critical Commentary," *Socialist Studies* 19, no. 1 (2025): 1–10.

30 Theodor W. Adorno, "The Essay as Form," *New German Critique* 32 (1958; Spring–Summer 1984): 159.

31 Michel de Montaigne, "To the Reader," in *The Complete Essays*, trans. M.A. Screech (1588; Penguin, 2003), lxiii.

32 Andrew Sullivan et al., "American Democracy Debate," Munk Debates, October 12, 2017, https://munkdebates.com/debates/american-democracy.

33 Sullivan et al., "American Democracy Debate."

34 Sullivan et al., "American Democracy Debate."

35 David Runciman, *How Democracy Ends* (Basic Books, 2018); Steven Levitsky and Daniel Ziblatt, *How Democracies Die* (Crown, 2018); Adam Przeworski, *Crises of Democracy* (Cambridge University Press, 2019).

36 Michelle Alexander, "We Are Not the Resistance," *New York Times*, September 21, 2018, https://www.nytimes.com/2018/09/21/opinion/sunday/resistance-kavanaugh-trump-protest.html.

37 Rosie Gray, "Trump Defends White-Nationalist Protesters: 'Some Very Fine People on Both Sides,'" *Atlantic*, August 15, 2017, https://www.theatlantic.com/politics/archive/2017/08/trump-defends-white-nationalist-protesters-some-very-fine-people-on-both-sides/537012/; Kathleen Ronayne and Michael Kunzelman, "Trump to Far-Right Extremists: 'Stand Back and Stand By,'" *AP News*, September 30, 2020, https://apnews.com/article/election-2020-joe-biden-race-and-ethnicity-donald-trump-chris-wallace-0b32339da25fbc9e8b7c7c7066a1db0f.

38 Corey Robin, "Trump and the Trapped Country," *New Yorker*, March 13, 2021, https://www.newyorker.com/news/our-columnists/trump-and-the-trapped-country.

39 Andrew O'Hehir, "One Election Can't Resolve the Crisis of Democracy," *AlterNet*, November 9, 2020, https://www.alternet.org/2020/11/one-election-cant-resolve-the-crisis-of-democracy.

40 Corey Robin, "A Month Before Election Day, I Predicted . . . ," *Corey Robin* (blog), December 21, 2024, https://coreyrobin.com/2024/12/21/a-month-before-election-day-i-predicted/.

41 Dylan Riley, interview by Suzi Weissman, host, *Jacobin Radio*, podcast, "Return of the Trumpian Right w/ Dylan Riley," Jacobin, November 18, 2024, https://podcasts.apple.com/ca/podcast/jacobin-radio-return-of

-the-trumpian-right-w-dylan-riley/id791564318?i=1000677332493;
Quinn Slobodian and Wendy Brown, interview by Daniel Denvir,
host, *The Dig*, podcast, "MAGA 2.0 w/ Quinn Slobodian & Wendy
Brown," Jacobin, November 29, 2024, https://thedigradio.com/podcast
/maga-w-quinn-slobodian-wendy-brown/.

42   Betsy Woodruff Swan and Kyle Cheney, "Trump Expressed Support
.   for Hanging Pence During Capitol Riot, Jan. 6 Panel Told," *Polit-
ico*, May 5, 2022, https://www.politico.com/news/2022/05/25/trump
-expressed-support-hanging-pence-capitol-riot-jan-6-00035117.

43   Andrew Coyne, "Trump's Election Is a Crisis Like No Other, Not
Only for the U.S. but the World," *Globe and Mail*, November 6, 2024,
https://www.theglobeandmail.com/opinion/article-trumps-election
-is-a-crisis-like-no-other-not-only-for-the-us-but-the/.

44   Graf and Jarausch, "'Crisis,'" 6.

45   Gabbert, "Epilogue," *Unreality of Memory*.

46   Nancy Fraser, "Legitimation Crisis? On the Political Contradictions
of Financialized Capitalism," *Critical Historical Studies* 2, no. 2 (2015):
186.

47   Fraser, "Legitimation Crisis?," 159.

48   Fraser, "Legitimation Crisis?," 167–75.

49   Antonio Gramsci, *Selections from the Prison Notebooks*, ed. and trans.
Quintin Hoare and Geoffrey Nowell-Smith (Lawrence & Wishart,
1971), 276.

50   G.W.F. Hegel, "Hegel's *Philosophy of Right* Preface," trans. S.W. Dyde
(1820; 1896; Marxists Internet Archive, n.d.), https://www.marxists
.org/reference/archive/hegel/works/pr/preface.htm.

51   Blanche Verlie, *Learning to Live with Climate Change: From Anxiety to
Transformation* (Routledge, 2022), 12.

52   David McNally, *Global Slump: The Economics and Politics of Crisis and
Resistance* (PM Press, 2011), 1.

53   Ernst Bloch, *The Principle of Hope*, trans. Neville Plaice, Stephen Plaice
and Paul Knight, vol. 3 (Basil Blackwell, 1986), 1374.

54   Franny Choi, "Doom," in *The World Keeps Ending, and the World Goes
On* (Ecco, 2023), 116.

THE WORSE, THE BETTER

Lord, I confess I want the clarity of the catastrophe but not the catastrophe.
– Franny Choi, "Catastrophe Is Next to Godliness"

The best time to teach classes about democracy is when mass movements are on the rise. When movements are thriving, students don't need my stories of the past to get a feel for the power of the people. They can look outside at Occupy Wall Street, the Quebec Student Strike, Idle No More, Black Lives Matter, strikes for climate justice and campus encampments for Palestine.

In 2011, the Government of Quebec announced plans to hike university tuition by 75 percent, no ifs, ands or buts, no negotiation. In 2012, a months-long student strike forced that government from office, effectively cancelling the hike. The power of policing seemed unshakable; then crowds in Minneapolis torched a cop station and millions around the world took to the streets while chanting, "Defund the police." I've watched students on my campus who've never thought of themselves as "political" organize radical reading groups linking themselves to Montreal, Minneapolis and more.

When I'm teaching at the height of social struggle not far away, I always ask students: What would it take for that to happen here? Montreal's not so far from my campus in Brantford, Ontario. For that matter, neither is Wet'suwet'en territory, or Oaxaca or Bengal. What would need to happen for movements surging elsewhere to spread? I have a suite of good answers in mind: solidarity actions, stronger city-to-city activist networks, bold leadership from big labour unions. But students never give these answers.

The answer I hear first and most often is this: "Things here would have to get a lot worse first." People only rise in their thousands and millions when life has gotten so bad that they have no choice but to resist. Every town has a left-wing fringe writing letters and demonstrating about every damn thing; but most people most of the time don't care enough to act.

Students don't frame the idea in theoretical terms, and I'm not (yet) pronouncing on its merits, but the idea that "things would have to get a lot worse" before mass unrest breaks out rests on a bold theory of social change. Core assumptions of the theory include: there is a direct correlation between increasing mass misery and increasing collective action; the status quo is not so bad, at least as measured by public opinion; and crisis is an essential component of social transformation. It must also assume that the historical record shows that when social conditions worsen, movements for social justice surge.

On the radical left, there's shorthand for this kind of thinking: namely, the-worse-the-better politics (hereafter TWTB). Some on the left build strategy from the idea. For example, the 2009 book *The Coming Insurrection* argues that small groups wishing to change the world need to *accelerate the contradictions* in society by committing political violence.

In class, I've always pushed back against TWTB thinking. I do so partly because it pains me to think that students' political horizons are so narrow that they can't conceive of weighty movements in political counter-flow developing locally. Partly, I want to challenge the idea that things in Brantford (or anywhere) are so good for most people. Mostly, though, I'm expressing my understanding of how change happens, which I've developed through years of (sporadic, modest) political activism and a great deal of reading and talking about radical social theory and history.

I say words to the effect of: But it isn't true, historically, that things only get better after they get worse. Paul Foot's *The Vote* shows that the UK's broad welfare state formed because of pressure from a working-class newly emboldened by defeating the Nazis. In postwar Britain, it was rising working-class confidence – a growing sense of collective entitlement to better lives – not worsening conditions that expanded social supports and deepened democracy. Certainly, we see victories-from-below inspiring new movements throughout the protest wave of the late 1960s. In *The Imagination of the New Left*, George Katsiaficas uses the concept of "the Eros effect" to refer to how people learn from, draw strength from, expand their sense of what's possible and take audacious new action when observing powerful campaigns by allied groups.[1]

I experienced what Katsiaficas was talking about when I marched among the half-million strong supporting the Quebec Student Strike in Montreal on May 22, 2012. Blocks and blocks of downtown were filled with marchers and their supporters waving from cafés and apartment buildings. Somewhere along Sherbrooke Street, not far from the main gates of McGill University, it struck me that the cops were not in charge of what would happen next. The people, in their overwhelming, shape-shifting mass, were in full control of the streets. I understood with my full body what John Berger was talking about in his essay on "The Nature of Mass Demonstrations." The true power of demonstrations, wrote Berger, isn't to appeal to the conscience of authorities (good thing, too, as the Government of Quebec had already declared countless times that under no condition would it revoke the tuition hike prompting the strike). Rather, the demonstration "*demonstrates* a force that is scarcely used": the collective strength of the subordinate classes.[2] The more people taking part in the demonstration, "the more powerful and immediate (visible, audible, tangible) a metaphor it becomes for their total collective strength." In short, progressive politics often do "better" as a result of doing better.

Then I make a corollary point: Sometimes things get worse, then just keep getting worse. When material conditions deteriorate, often people give up – too broken to fight back, too cynical to organize. In the days following Trump's 2024 election victory, the Princeton historian and socialist activist Keeanga-Yamahtta Taylor describes precisely this dynamic. Trump's first presidency was dangerous and disgusting, says Taylor, but it was President Biden who dismantled the Covid welfare state, Biden who funded and provided diplomatic cover for the Israeli genocide in Gaza, and Kamala Harris who campaigned from the right on everything from the border to fracking to the need to appease Republican elites. The economic and political situation of working-class people deteriorated under Biden-Harris; then Trump was elected again, and things went from bad to worse. "There's this idea from liberals," says Taylor, "that, 'Oh, this has happened before, we confronted this before, let's have a Women's March in January . . .' And, things are worse! Things are materially worse for people," and this time Trump's governing team has a policy plan they didn't have in 2016.[3] "They have the government! They have all the chambers of Congress. They have the Supreme Court! This is worse!" The brutality of Trump's first administration, the horrors of Biden-Harris and now? Trump 2.0.

History is full of situations that couldn't be much "worse" when collective action was rare or non-existent for decades. European colonization of North America made life worse for Indigenous people. It got worse, then worse. The terror of the Nazi regime didn't trigger a mass coordinated fight back among Jews in Europe. The transatlantic slave system held for more than three hundred years. I'm not ignoring the many examples of resistance to these brutal regimes: the Warsaw Ghetto Uprising, the Red River Rebellion, the Amistad mutiny, righteous assassinations of slave owners. I'm saying that "the worse, the better" theory of change runs into trouble when it comes to historical periods when worse led to worse. As for the motivating power of bad news, Ezra Markowitz and Lucia Graves note that "crisis and catastrophe narratives don't keep people engaged for long, especially when there's a new emerging crisis to worry about every few weeks or months."[4]

I think of myself as belonging to the same political tradition as Ståle Holgersen, who criticizes the tendency among some on the left to welcome the outbreak of crisis, as though the next disaster could be the one to bring the ruling class to its knees. Making things worse will not build people power. I think of myself as the kind of radical who says, "What we need is to build people's confidence through winning reforms, showing that change is possible."

And yet.

And yet.

It feels as though every far-left meeting I've ever been to begins with a speaker detailing just how bad things are today and forecasting much worse ahead. The socialist press invariably dubs each economic crisis the worst in history. My social media are filled with political radicals giddy over negative economic forecasts and the latest promises broken by centrist governments. Two weeks after Trump's second election win, a socialist collective in the US organized a panel under the proposition: "Trump is back, Fascism is back on the horizon."[5] There's no denying it: aspects of TWTB politics are embraced even by lefties who wouldn't put it that way. What's going on? Does the path to liberation run through catastrophe? If so, why do some intellectuals and activists, myself included, have such a hard time admitting that? If not, why do we on the far left revel in bad news?

In a loud Manhattan diner in spring 2011, my Marx-ish friend smirked as she said it: "Marxists love a crisis." We'd just seen a panel at the Left Forum at Pace University, where two men named David analyzed the 2008–9 economic crash. The Davids debated whether the crash was a crisis of underconsumption, overproduction or both. They described stages through which the crisis had morphed (from housing to banks to sovereign states), and speculated about the struggles to come. Zuccotti Park is just around the corner from Pace University, but the birth of Occupy Wall Street was still six months away.

I'd been active in a revolutionary socialist organization for three years at that point.[6] In fact, it was the 2008 crash that forced me to get serious about radical politics. Through graduate school in the early 2000s, I thought reading French postmodernists, growing a rat-tail and wearing puffy blouses would foment revolution. Okay, I didn't really think that. But I did have a foot-long rat-tail. And I did wear a frilly pirate shirt under a cape to several academic conferences. I half-believed that unconventional dress exposed the artifice of bourgeois society. How is my cape any more ridiculous a uniform than your Oxford shirt and khaki pants? Both are just fabric. The world can mean whatever we want it to.

Then, over a matter of days in October 2008, events with very concrete meanings began to unfold. Stock markets crashed. Banks threw millions of people out of their homes. Factories shuttered. Governments who all my life had said that there's nothing to do but cut, cut, cut, now shovelled trillions of dollars into private companies. Workers, too, were lashing out in ways I'd not seen before at such scale or with such militancy: kidnapping bosses in France, occupying factories in the US, waging a mass revolt in Guadeloupe and Martinique.

There were weeks in the fall of 2008 when I honestly wondered whether the system would survive. While US Secretary of the Treasury Hank Paulson was admitting to his wife, "I am really scared,"[7] I was sitting on an overturned milk carton in my Toronto apartment, reading articles in the pro-business *Globe and Mail* about whether this was the end of capitalism.

I recall saying to my partner at the time: "What the fuck am I doing?" She was lying on the couch we'd pulled from the curb into our living room. "Social order is collapsing, and I'm in the basement writing about what the governor general wore to the 1905 Throne Speech?" (That is, actually, more or less, the topic of my PhD dissertation.) In that precise moment, under the glass eyes of the taxidermied fox we'd propped up next to our television, I renounced my dalliance with postmodernism, and pledged myself to radically democratic struggles for social justice. Even if being a revolutionary socialist turned out to be as boring and ineffectual as I'd feared it would, it would still be better than shutting out the world through the intellectualized nihilism I'd been cultivating.

I regret that it took me until the age of twenty-nine to choose activism over stylized apathy. I also know that not everyone changes political course while living through a social crisis. I tell my story of personal transformation because it gets at the wisdom in my friend's teasing about Marxists loving crises. Marxists love crises not primarily as means of theoretical point-scoring, or because they take pleasure in watching class enemies suffer (though I like watching bosses kidnapped as much as the next commie). The fondness for crisis goes to the Greek root of the word, the verb *krino/krinein*, meaning to separate, choose, decide, judge. The Italian Marxist Antonio Gramsci once said that "living means taking sides."[8] Because crises tend to lay bare real antagonisms typically hidden below the surface in "normal" periods, they are times of choosing sides; for Gramsci, times of living. By exposing and expressing the violence and instability of social order, crises are openings toward the potential of a better future.

When Covid-19 kicked off another worst-ever capitalist crisis in March 2020, there I was in *Spring* magazine, writing: "As the crisis exposes dirty secrets of capitalism normally hidden under the laws and rhetoric of business-as-usual, a window has been flung open through which it's easier to see not simply the irrationality and brutality of capitalism, but the fact that market rule is not the only way to organize society."[9] Am I more of a the-worse-the-better guy than I realized?

It's impossible to imagine today's growth in movements for ecological justice without the worsening ecological conditions of the past decade. Recent successful union drives in hyper-precarious work sectors (like those at Starbucks and Amazon) are direct responses to worsening conditions of work in the gig economy, unaffordable housing and soaring inflation. The re-emergence of socialism in the Anglo-American mainstream since 2008 has taken place against the backdrop of two once-in-a-lifetime global crises of capitalism, growing inequality and the dismantling of the welfare state, which is, in short, to say, working-class loss. Set aside for a moment the question of whether progress in left-wing movements outweighs the hardship of worsening conditions. If one goal of radicals is to learn from history to inform our efforts to change the world, then let's be honest that sometimes worse leads to better.

The phrase itself is often attributed to the Russian revolutionary Vladimir Lenin. In a *Christian Science Monitor* piece from 1980, New York University professor Albert L. Weeks wrote: "'Worse is better' was [Lenin's] slogan. Destruction-followed-by-construction was the Leninist sequence."[10] (Weeks conveniently ignores the fact that he could just as easily have been describing that free-market champion Joseph Schumpeter, who, in 1942, lauded the "creative destruction" of capital.[11]) A 2008 article in *Forbes* referred to Lenin's "old 'the worse, the better' ploy. [. . .] What he meant is that the worse things go, under the czar or during the chaos following the czar's toppling, the greater the chance that ordinary Russians would turn to Lenin's brand of Marxist revolution."[12] In 2020, Walter Clemens, a Harvard professor, wrote that Lenin and Marx would've loved Donald Trump because "Leninists don't care how bad things get. For them, 'the worse, the better.'"[13]

If Lenin had said "the worse, the better," he would've been

partially correct in the context of Imperial Russia. The Tsar's authoritarianism and warmongering no doubt drove many Russians toward revolutionary politics. This isn't the same as saying that Lenin loved seeing peasants starved and comrades executed by the state. However, every history of the Russian Revolution (Trotsky's towering book included) frames Tsarist terror as a major reason why democratic political parties ballooned in the early years of the twentieth century. In a similar vein, you could say (many have) that the Jallianwala Bagh Massacre of 1919 caused Gandhi's commitment to national liberation. Police raids triggered the Stonewall Uprising of 1969, the watershed of the gay rights movement. Many friends and comrades have told me how they suddenly saw the world in new ways as a result of being evicted, or being attacked by the cops or battling through a long strike.

But here's the thing: Lenin didn't say the worse, the better. Okay, he did write those words in a 1901 article in the political journal *Zarya*. But he wrote them with the express purpose of repudiating the general principle! At the time, Lenin was criticizing the reformist wing of European socialism, which believed that communism could replace capitalism by gradual, step-by-step changes to government policy. Lenin condemned the "narrow-mindedness and stupidity" of reformists who believed, as a rule, that "the better things are, the better."[14] This general principle ignores the history of ruling powers making concessions to opponents in order "to disunite the attacking party and thus to defeat it more easily." However, Lenin also rejects the simple inverse of reformism. In his words, "This principle [of the better things are, the better} in its general form *is as untrue as its reverse* that the worse things are, the better'" (italics added).

Only a few years after Lenin's article appeared, the Tsar's violent crackdown against political dissent following the Revolution of 1905 effectively crushed democratic organizing in Russia for nearly a decade. In that sense, worse (from the Tsar) wasn't better (for the revolutionaries). Moreover, you might say that the 1906 "better" (better at least in the eyes of liberal reformists), in the form of the Tsar's modest political reforms was worse for Lenin and the far left. Reform broke the fragile unity among the Tsar's political opponents. Liberals and social democrats alike channelled energy into the new Duma, a severely limited, ultimately impotent representative institution.

We're now faced with a jumble of worses and betters pointing in so many different directions that it's pointless to try to formulate a universally valid, coherent theoretical conclusion. (And we've said nothing about the plight of revolutionaries during the First World War, the formation of the soviets in 1917 or the civil war – yet more contradictory examples of worse-and-better in each moment.) You'd find the same jumble no matter what episode of historical struggle you examined. Your perplexity would only deepen the more thoroughly you researched. The women's movement, battles for national liberation, union struggles, the civil rights movement, struggles of queer people – all are records of harms endured, gains made, losses and victories (some more partial than others), that include moments in which worse appears to lead to better, other moments demonstrating the opposite relationship and yet more moments in which the only sure dynamic is contradiction.

Returning to Lenin has led me to realize that asking if worse is better is the wrong question. I've been drawn into the debate countless times with students, comrades and by my own inner critic. But debate, no matter how thorough the evidence, no matter how long the argument, cannot settle the dispute. Sometimes worse is better; sometimes it isn't. The point is summed up nicely in (anti-Communist) Thomas Hammond's scholarly study of Lenin's view of Russian trade unions:

> [Lenin] did say that a worsening of conditions, as in a depression, might prove to some of the workers that revolution was the only

way. On the other hand, he also pointed out that under proper circumstances the winning of improvements (which he believed could be only temporary) might be accompanied by a strengthening of revolutionary spirit and an increase in mass participation in the struggle against capitalism. Sometimes the achievement of temporary economic reforms by the workers resulted in a "corruption of political consciousness," but on other occasions it strengthened class solidarity. Consequently, from Lenin's viewpoint it would be impossible to make a broad generalization either that "the worse things are the better," or that "the better things are, the better."[15]

When movements for social justice and ecological sustainability don't appear to be winning, it's understandable why it seems to some that the only way for things to improve is for everything to fall apart first. The problem, for one thing, is that worsening social and ecological conditions entail worlds of pain at the individual level. It's sadistic (and movement-defeating) to want our neighbours to suffer. Second, believing that worse is the source of our salvation effectively nullifies the claim to want better. If worse is better, then join shills for the oil and gas industry chanting, "Drill, baby, drill!"[16] Vote for racists and warmongers. Launch a union decertification campaign at your workplace. Help Stephen Miller implement Project 2025. How much worse does it have to get before it's okay to start building again? To those inclined to think that broadcasting worsening pain spurs people into action, the anthropologist Joseph Masco says that "crisis talk without the commitment to revolution becomes counterrevolutionary."[17]

Most important, though, when people who want social and environmental justice believe TWTB, they ignore the reality that there is potential for movement advancement and risk of movement losses – promise and threats, better and worse – in every moment, every turn of events, every rebalancing of power. The key question for radicals and students of revolution isn't whether movements are helped more by social improvements or hardships. The question in every situation, worse and better, is: Which forces are feeding revolutionary struggle, and which forces are defending current systems of inequality, suppressing democracy-from-below? In Lenin's words, "Only by having the 'ultimate aim' in view, only by appraising every 'movement' and every reform from the point of view of the general revolutionary struggle," is it possible to develop sound strategy.[18] This means seeing both potential gains and losses in every worse and every better. In the case of the 2024 US election, rejecting simplistic notions of worse (as Trump) and better (as Harris) led some US-based progressives to refuse to support Harris's candidacy. For example, the Palestinian-American lawyer and professor Noura Erakat, wrote on social media that voting for Harris on the grounds that it's a vote against fascism

> imagines that fascism is emerging only from a Trump camp. It does not recognize how the mobilization of state violence, identity politics, the disregard of law/accountability, rejection of 'truth,' media censorship, & normalization of genocide stems from the Dems now.
>
> It also forgets that in 2020, this was the same impulse that united the left with the Dems to defeat Trump. The compromise then was dropping abolition as a serious social and political priority.
>
> With it, dropped a critique of how securitizing borders, neighborhoods, and ppl served to protect an economic & political elite that continued to accumulate and hoard resources through dispossession. The party has since platformed police and delivered genocide.[19]

When we look from the point of view of radical struggle, we're able to see suffering driving collective action and suffering leading to apathy. We observe political victories leading to bold experiments in struggle (as the victory against US imperialism in Vietnam fed struggles for decolonization around the world) and political victories pacifying movements (as legal protections around collective bargaining – huge, essential wins – played a role in demobilizing the union movement in the second half of the twentieth century). To say that worse (in the form of the 75 percent Quebec tuition hike) is what made things better (in the form of a militant, popular strike and rolling mass demonstrations, what Berger would call "rehearsals for revolution"[20]) is to miss the worse and better that went into the confidence-building organization of popular assemblies across campuses and decades of student organizing in Quebec prior to 2011–12. Worse: both cause and effect; better: both cause and effect.

From the point of view of the general struggle for social and ecological justice, crises are neither wholly good nor wholly bad. They are, like all political episodes, a collection of forces fuelling resistance and reaction. They are uniquely condensed moments of struggle – condensed both in terms of time (they are brief), and in terms of significance (the stakes are high). But they don't possess an inherent worseness or betterness outside of the context of past and future struggles. The primary political question is not what crises *do to* people, but what people do with crises when crises break out.

My friend in the Manhattan diner was partially right when saying that "Marxists love a crisis." But only to the extent that crises unleash new possibilities for progressive collective action. Look around during a crisis: What are the most advanced, organized sections of the radical left doing? Fighting against layoffs, campaigning for "a people's bailout," organizing disaster relief hubs, protesting war, supporting struggles of undocumented people. They're not making merry in the ruins of the crisis but trying to protect people from the worst of its effects. And when gains are made, during crises and in calmer times, activists celebrate what's been won, publicize the victory, strive to create momentum that could carry onto new ground.

If anything, Marxists are guilty of exaggerating the extent of working-class advances. The very same radicals who emphasize the depth of the crisis also portray every new episode of mass mobilization as the spark that could set off the fire next time. If Marxists love a crisis, they also love workers winning job security, the defeat of right-wing street gangs, periods of full employment (because they improve workers' bargaining position). Answering my question of why leftists revel in bad news involves answering the question of why leftists revel in good news. They're considering events from "the point of view of the general revolutionary struggle."[21] In every worse and better, there are forces fuelling struggle-from-below, and forces fuelling reaction. Radicals feed the struggle and fight the reactions. Crises are worse and better; periods between crisis are worse and better.

NOTES

1 George Katsiaficas, *The Imagination of the New Left: A Global Analysis of 1968* (South End Press, 1987), 7, 10–11.

2 John Berger, "The Nature of Mass Demonstrations," *International Socialism* 32 (Autumn 1968): 11–12, italics in original.

3 Keeanga-Yamahtta Taylor, interview by Daniel Denvir, *The Dig*, podcast, "Democratic Dealignment w/ Keeanga-Yamahtta Taylor," Jacobin, November 9, 2024, https://thedigradio.com/podcast/democratic -dealignment-w-keeanga-yamahtta-taylor/.

4 Ezra Markowitz and Lucia Graves, "After 2020, We Need to Talk About How We Talk About Catastrophe," *Washington Post*, December 30, 2020, https://www.washingtonpost.com/outlook/2020/12/30/crisis -catastrophe-communications-action/.

5 Tempest (@tempest_mag), "Trump is back – emboldening the return of fascist forces. So how do we fight it?," X, November 13, 2024, https://x.com/tempest_mag/status/1856768568752521305.

6 In contrast with versions of socialism that believe that a socialist soci-

ety can be achieved within current political-economic structures, revolutionary socialism assumes that achieving social and environmental justice requires a fundamental break with (or overthrowing of) capitalism and the state.

7 Quoted in David McNally, *Global Slump: The Economics and Politics of Crisis and Resistance* (PM Press, 2011), 13.

8 Quoted in Alan Wald, "Gramsci's Gift," *Boston Review*, April 4, 2022, https://www.bostonreview.net/articles/gramscis-gift/.

9 James Cairns, "There Is an Alternative," *Spring*, March 30, 2020, https://springmag.ca/there-is-an-alternative.

10 Albert L. Weeks, "The Long Shadow of Lenin's 'Worse Is Better,'" *Christian Science Monitor*, April 23, 1980, https://www.csmonitor.com/1980/0423/042334.html.

11 Joseph Schumpeter, *Capitalism, Socialism, and Democracy* (1942; Routledge, 1994), 139.

12 Rich Karlgaard, "The Worse, the Better," *Forbes*, December 26, 2008, https://www.forbes.com/sites/digitalrules/2008/12/26/the-worse-the-better/.

13 Walter Clemens, "Marx and Lenin Take Washington," *Russia and America in the 21st Century* 2020, no. 1 (2020): para. 9.

14 V.I. Lenin, "The Persecutors of the Zemstvo and the Hannibals of Liberalism" (*Zarya*, 1901; Marxists Internet Archive, 2003), https://www.marxists.org/archive/lenin/works/1901/jun/15.htm.

15 Thomas T. Hammond, "Lenin on Russian Trade Unions Under Capitalism, 1894–1904," *American Slavic and East European Review* 8, no. 4 (1949): 287.

16 See *Post* Editorial Board, "Drill, Baby, Drill: Cut off Putin's Power by Cutting Off Russian Gas," *New York Post*, February 28, 2022, https://nypost.com/2022/02/28/cut-off-putins-power-by-cutting-ourselves-off-russias-gas/.

17 Joseph Masco, "The Crisis in Crisis," *Current Anthropology* 58, no. S15 (2017): S73.

18 Lenin, "Persecutors of the Zemstvo."

19 Noura Erakat (@4noura), X, October 25, 2024, https://x.com/4noura/status/1849881475102933385.

20 Berger, "Nature of Mass Demonstrations."

21 Lenin, "Persecutors of the Zemstvo."

ON THE PLEASURES OF READING THE APOCALYPSE

> What is sung by the prophets is but the same song sung across time, the coming of the sword, the world devoured by fire, the sun gone down into the earth at noon and the world cast in darkness.
> – Paul Lynch, *Prophet Song*

How many times have novels and movies shown me the end of the world? I watch humanity wiped out by nuclear war, or an asteroid, or a plague, or a flood. Survivors hunt for and hide from one another in wrecked buildings and toxic swamps.

We go to end-of-the-world fiction for two obvious reasons. First, we want distraction. Explosions onscreen can block out explosions in our lives. I'd rather worry about storms in the movie *The Day After Tomorrow* than the tasks I said I'd finish before actual tomorrow.

Second, perhaps incongruously, we want to feel hopeful. In Octavia Butler's *Parable of the Sower*, civilization is collapsing, yet Lauren Olamina never wavers from her commitment to survival and rebirth. At the end of Waubgeshig Rice's *Moon of the Crusted Snow*, the Anishinaabe community leaves its apocalypse-ravaged reservation for a new beginning in the woods. In *Armageddon*, Bruce Willis blows up Michael Bay's asteroid moments before Armageddon commences. Even Cormac McCarthy's "emotionally shattering" *The Road* ends with the adoption of the newly-orphaned boy in the wake of his dead father's command to go on.[1] The moral of the stories: We, humanity, shall overcome. In a *Globe and Mail* piece, David Moscrop wrote that while reading apocalyptic novels might not be "comforting," it can encourage "valuable reflection about how we live our lives today – and how we might choose to live them tomorrow."[2]

Rumaan Alam's apocalyptic novel *Leave the World Behind* (a 2020 finalist for the National Book Award) enchants for a different reason. By painting a picture of total human annihilation – no plucky survivors, no one spared by design or by chance – the book offers the relief of surrender.

Alam's novel begins with a white, middle-class family arriving at a bucolic vacation home east of New York City. The family splashes in the pool and fantasizes about owning marble countertops, solid oak floors, ample space. The mom, Amanda, can't resist checking her work email. Clay, the dad, sneaks cigarettes in the driveway while cooking up half-baked media studies theories to test out on his fall crop of college students. The kids – Rose, ten, and Archie, thirteen – look at their phones.

The centrality of technology is true to life and crucial to the plot. Cell signals, the internet and cable television stop working shortly after the family lands in the countryside. Probably, they think, their remote vacation spot is beyond reach of satellite networks. That night, though, when the owners of the house, the Washingtons, a kind, elderly Black couple, show up and ask to stay, Clay and Amanda learn that the loss of service is widespread. The power is out across New York City.

Drama unfolds on two tracks. There is tension between the families. Clay and Amanda are suspicious of the Washingtons, which has as much to do with the white couple's latent racism as with the unexpected appearance of the homeowners. Who has the right to call the shots: the white renters or the Black deed holders? At what point does valid speculation about the crisis slide into harmful paranoia?

On a second narrative track, the world is ending. The reader understands this early in the book more clearly than the characters ever do. There's plenty of evidence on Long Island that something is wrong. The blackout, communication breakdown, a deafening noise overhead, terrified neighbours, flamingos in the pool. A few days after the vacation begins, Archie's teeth fall out. The families know there is trouble, they are in trouble, but they never understand the extent of it. Not knowing is part of their terror.

By contrast, readers know that we're witnessing the end of the world because Alam periodically interrupts action in the vacation home with a Voice of God narrator (VOG) describing cataclysmic events far off. While Clay and Amanda fuss over the lost TV signal, the VOG tells us: "A storm had metastasized into something for which no noun yet existed [. . .] the electrical grid broke apart like something built of Lego [. . .] machines meant for supporting life ceased doing that hard work after the failure of backup generators in Miami, in Atlanta, in Charlotte, in Annapolis."[3]

Around the novel's midpoint, a horrifying noise erupts from the sky. The noise divides the families' lives in two: "the period before they'd heard the noise and the period after."[4] Inside the novel, no one discovers the source of the sound. However, readers learn from the VOG that top-secret fighter jets are scrambling toward a new era of battle over the eastern seaboard.

Far from New York, the VOG explains, "some people were committing suicide. Some people were packing things up in cars and hoping they'd be able to get a mile or two or ten or whatever it would take to reach wherever safety endured. [. . .] Some people didn't know anything was amiss. [. . .] Some people started to realize they'd had a naive faith in the system. Some people tried to maintain that system. Some people were vindicated that they'd stockpiled guns and those filter straws that made any water safe to drink. However much had happened, so much more would happen."[5]

If there were no VOG interruptions, no recurring omniscient assurances anchoring the contingencies of the interpersonal plot to the certainty of global apocalypse, *Leave the World Behind* would be an anxiety novel. Is Armageddon nigh or not? Some of my favourite books are anxiety novels. I identify with the catastrophism and fear expressed in books like Ben Lerner's *10:04* and Jenny Offill's *Weather.* In fact, I find it comforting that the anxiety of *Weather*'s narrator (Lizzie) far surpasses my own. Watching Lizzie worry as she does affirms the validity of my worries and tells me that my anxiety isn't nearly as bad as some people's.

Arguably, the end-of-the-world anxiety novel is scarier than speculative end-of-the-world fiction. The anxiety novel is about the world as it is: the painful truth of real existence, not the threat of an imagined existence-to-come. Anxiety is torturous, paralyzing. It's a truism of the horror genre that anticipating the arrival of the monster can be more terrifying than the beast's appearance.

But the uncertainty driving the anxiety novel, the book's ultimate source of terror, can't help but leave open the possibility that things might not be as bad as they seem. Notwithstanding the disastrous portents of the present, the future could turn out better than we'd predicted (especially given that anxiety, even well-founded anxiety, always contains traces of paranoia). No matter how implausible the idea of post-novel happiness is in *Weather,* it is conceivable that movements for eco-sustainability could usher in a world less destructive than Lizzie imagines. The narrator in Lerner's *10:04* pictures today's superstorms as tomorrow's ice age . . . but, then, maybe they won't be.

In *Leave the World Behind,* there is no uncertainty. When Amanda wonders if war has erupted suddenly, the VOG interrupts: "She didn't know that it was worse, that war could not describe it," and that there is nothing anyone can do about it.[6] Apocalypse is underway, whether you can see that, worry about it, resist it or not.

Because if the bombs are already in the air, the electrical grid is already down for the final time, the life-destroying echoes of *the noise* are already in your body, there is no future that isn't mass slaughter. As if to put a fine point on the guarantee of imminent death, the futility of resistance, Alam bores an unnoticed tick into Archie's ankle before the boy is dying from noise-sickness.

In spring 2022, when I read *Leave the World Behind* with an undergraduate seminar called "The Crisis Today," several students saw signs of hope in the book's final chapter. Rose, the young girl, goes into the woods alone to find a house she'd seen on an earlier hike. No one answers when she knocks on the door. The VOG tells us that its occupants, stranded in California, "would never see this house again in their lives, though Nadine, the matriarch, would sometimes dream of it before she succumbed to cancer in one of the tent camps the army managed to erect outside the airport. They'd burn her body, before they stopped bothering with that, as the bodies outnumbered the people left to do the burning."[7]

Brave Rose breaks in to discover what first might seem like a lifeline. Food, supplies, living space. A place to wait out the storm? A new beginning? In a different novel, yes. Butler's Lauren Olamina could've turned the place into a plush Earthseed settlement. But the apocalyptic prognosis Alam has developed by this point in the book forecloses the possibility of a future here.

The vision of safety in the house is a mirage; worse, the illusion of hope mocks Rose's hopeless situation. Knowing what we know about the collapse-in-progress, the things in the house that symbolize safety to Rose – a bit of food, soft carpeting, camping gear – are no more promising than her dead phone. A giant TV that doesn't work (none do), a jar of pickles, a bottle of Advil, a few batteries: this stuff can't save them. Nothing can.

Why does Alam's crushing story captivate me? Why am I thrilled by the promise that we're on the edge of extinction?

I think the book delights by allowing us to revel in the pleasures of giving up. Quit your job, break dinner plans, stop exercising, leave the relationship. What joy there is in not having to do the thing we thought we had to do. The world is ending *and there's absolutely nothing you can do about it.*

In his essay "On Giving Up," the psychoanalyst Adam Phillips writes: "We tend to think of giving up, in the ordinary way, as a lack of courage, as an improper or embarrassing orientation toward what is shameful and fearful. That is to say we tend to value, and even idealise, the idea of seeing things through, of finishing things rather than abandoning them."[8] However, Phillips argues, there is such a thing as "a tyranny of completion, of finishing things, which can narrow our minds unduly." The refusal to give up can be harmful, murderous. Phillips interprets *Macbeth, King Lear* and *Othello* as tragic dramatizations of the tyranny of completion.

He could've mentioned capitalist ideology's emphasis on resilience, grit and bootstrap-pulling. In the eyes of employers and the state, working people must never give up striving. No matter your low wages, your skyrocketing rent, you must persevere, try harder. To give up can be to walkout, strike, object. There's radical political potential in Phillips's suggestion that "we should not underestimate the pleasures of giving up, however forbidden or shameful they may seem to be."

My earliest memory of the desire to give up ends with my mother rejecting it. (If Phillips were my therapist, no doubt I'd be encouraged to reflect on this memory to better understand my attachment to Alam's novel.) I was nine or ten years old and wanted to quit the school choir. Mom and I stood in the kitchen before breakfast. I don't remember why it felt so important to quit, but I was crying, shaking, desperate for the relief of not having to sing that afternoon.

Mom's response was sympathetic but stern: No. We don't quit things partway through. It's fine to not want to do choir. You don't have to be in the choir next year. But you made a commitment, and you will see it through. No negotiation. I felt like puking. How could Mom so strictly enforce the principle that we don't quit things when she'd just left her marriage of eighteen years? (And now that I've asked the question, perhaps I've also answered it.)

I have quit things, though. And I've loved it. In the second year of my undergraduate studies, I dropped a statistics class late in the term. Because I'd missed the course drop date, I was required to get the professor's signature granting me special permission. The prof's office was hot, and I was wearing a parka. Sun streamed in the window behind his head. I thought of how in class once he'd teared up while talking about the pain of losing his wife to cancer. I started justifying my decision to drop the course in the direction of his silhouette. He took the paper from me and signed it while I rambled. "People do things for different reasons," he said. "It's fine." The relief I felt far exceeded the joy produced by any A-plus I ever got.

Oh, the joy of leaving that troubled ten-year relationship! I imagine it's what Scrooge felt waking on Christmas morning, learning that he has another chance. I instantly recall the butterflies, the excitement of quitting what seemed like a life destined for permanent frustration. The breakup was terrible. I hated hurting her. The logistics of moving were complicated, and she trashed the house when she left the final time. But I don't feel the pain of those hurtful memories as intensely as I feel the pleasure of the memory of giving up.

Several students in my crisis seminar dismissed Alam's novel for being little more than crisis-porn for rich white people who've never experienced an actual crisis. In the foreword to *Catastrophism: The Apocalyptic Politics of Collapse and Rebirth*, Doug Henwood dismisses the notion that social movements will grow only after more people grasp how truly catastrophic our times are. "Catastrophe can be paralyzing, not mobilizing. Revolutionaries should be talking about possibilities of transformation, not spinning tales of great chaos and suffering."[9] It's possible that acclaim for Alam's novel stems from the nihilism of those who've had it easy.

However, considering the pressures on individuals to strive, take your licks, work harder, be thankful for what you have, it wouldn't surprise me if rich white people weren't the only ones who fantasize about giving up. Saddled by debt? That's your fault. Can't find work in your field? You should've studied something employers actually want. Trouble with immigration authorities? You're the one who chose to come here.

The incredible thing is that most of the time, people don't give up. They struggle, they overcome, they get by, they make do. Why don't people kill themselves, asks Camus at the start of *The Myth of Sisyphus*.[10] Life is absurd; what's the point of living? Many people feel as though life is unfair: full of back-breaking labour doing unfulfilling, poorly compensated tasks. Notwithstanding its obviousness, Camus's conclusion is profound: the nature of the human condition is to keep going, to *not* give up. It is human to rebel against the absurd and material obstacles to survival. That doesn't mean we don't fantasize about quitting, maybe even about leaving the world behind.

It's the pleasure in the dream of quitting, the fantasy of the feeling of the void, not the politics of mass death, that I desire. In imagining the end of the world, I experience the release of countless other pressures. My own anxieties get transferred to the novel, where they disappear, if only for a fraction of a moment, in the blackout, the sound, the carnage of the plot. This tendency of mine, which first struck me as troubling (am I really so sadistic?), might even be grounding, reinvigorating, healthy. Easing chronic stress and fear is a good thing. It's why some people meditate or jog. Research shows that watching horror movies can relieve psychological tension.

There are better apocalyptic novels than *Leave the World Behind*. For portraying social collapse as gradual and incomplete, Butler's *Parable of the Sower* and Emily St. John Mandel's *Station Eleven* are doubtless more realistic depictions of how modern society falls apart. Impoverished and homeless, surrounded by violence, drought and disease, Butler's Lauren Olamina has allies helping to establish a community rooted in the revolutionary vision of God-as-change. Mandel's roving theatre troupe carries the nourishing power of art over the pandemic-ravaged deadlands of northern Michigan.

The spirit of those books reminds me of Andreas Malm's admonition to fight climate change no matter the chances of victory. In *How to Blow Up a Pipeline*, Malm argues that even if we know for certain that the climate crisis cannot be stopped there remains a moral imperative – a species-defining need – to fight until our last breath. "If it is too late for resistance to be waged within a calculus of immediate utility, the time has come for it to vindicate the fundamental values of life, even if it only means crying out to the heavens."[11] Malm reveres the Jews in ghettos and camps who rebelled against the Nazis, knowing always they would die. "Better to die blowing up a pipeline than to burn impassively," writes Malm. The words could've come from Lauren Olamina's mouth.

In Rice's *Moon of the Crusted Snow*, once it's clear that widespread disaster has struck in "the south" (the heartland of Canada, and, presumably, the world), Aileen, a community elder, says to her neighbour, Evan:

"You know, when young people come over, sometimes some of them talk about the end of the world. [. . .]

"They say that this is the end of the world. The power's out and we've run out of gas and no one's come up from down south. They say the food is running out and that we're in danger. There's a word they say too – ah . . . pock . . . ah . . ."

[Evan:] "Apocalypse?"

"Yes, apocalypse! What a silly word. I can tell you there's no word like that in Ojibwe. Well, I never heard a word like that from my elders anyway. [. . .]

"The world isn't ending [. . .] Our world isn't ending. It already ended. It ended when the Zhaagnaash [white man] came into our original home down south on that bay and took it from us. That was our world. When the Zhaagnaash cut down all the trees and fished all the fish and forced us out of there, that's when our world ended. They made us come all the way up here. This is not our homeland! But we had to adapt and luckily we already knew how to hunt and live on the land. We learned to live here."

[. . .] "But then they followed us up here and started taking our children away from us! That's when our world ended again. And that wasn't the last time. We've seen what this . . . what's the word again?"

[Evan:] "Apocalypse."

"Yes, apocalypse. We've had that over and over. But we always survived. We're still here. And we'll still be here, even if the power and the radios don't come back on and we never see any white people ever again."[12]

Aileen (and Butler and Mandel, and every book and movie portraying post-apocalyptic human struggle) is very likely right in assuming that the world will not end all at once. In *Station Eleven*, twenty years after the pandemic killed 99.99 percent of the human species, characters refer to themselves as living in the world after the end of the world. Aileen's voice echoes in Franny Choi's poetry: "By the time the apocalypse began, the world had already ended. It ended every day for a century or two."[13] In the final pages of *Prophet Song*, Paul Lynch writes that "the world is always ending over and over again in one place but not another and that the end of the world is always a local event."[14] Viewed in one light, the world will not end even if it does.

Of course, in a different light, one capable of simultaneously illuminating past, present and future the world will end, *is ending*. It's just a matter of time. In an essay about art's ability to alter experiences of time, Karl Ove Knausgaard writes: "We see the changes in the clouds but not the changes in the mountains," because the "now" of human perception excludes geologic time.[15] In reality, mountains are moving, just more slowly than rivers and rabbits. Humans may hold on for another few million years (although the smart money isn't betting on it); but ultimately, the existence of our species, our planet, is finite. It's anyone's guess how life on earth is eventually snuffed out for good. Fire? Ice? Alien invasion? In any case, the party won't last forever.

Butler's and Mandel's realistic depictions of the gradual, uneven nature of collapse can make Alam's Big Bang version of the final crisis look foolish by comparison. But Alam is not wrong that one day it will all end in the passage of one second to the next. The light will be on, as it has been for millennia, and then the light will go out. Alam's innovation is drawing that uniquely decisive moment from the (hopefully far-off) future and placing it in the now. Lights out tomorrow or next week. Or after lunch.

Whereas Butler's, Mandel's and Rice's main characters brim with insights about societal change and social justice, Alam's self-absorbed middle-class cast lusts over money and searches for Coca-Cola. Yet while I identified more closely with the politics of *Station Eleven*, *Moon of the Crusted Snow* and *Parable of the Sower*, and enjoyed those books immensely (few would dispute that Butler's book is a major contribution to late twentieth-century literature), none enthrall me as does *Leave the World Behind*. Stories of reproducing lives and communities in the aftermath of civilizational collapse are inspiring, admirable and satisfying. They're also exhausting, and not only because there are fires to build, continents to trudge across and gangs of murderous thieves to avoid. There's also the intense, inescapable fear on every page that survival won't work out. Nothing is guaranteed in the post-apocalyptic novel of endurance. By contrast, Alam's book guarantees the sudden and utter end of it all. There's catharsis in the swiftness and totality of such destruction.

The *Oxford English Dictionary* defines "catharsis" as "the purification of the emotions by vicarious experience."[16] Philosophers since Aristotle have praised the cathartic potential of art. Amid today's overlapping political, economic and ecological crises, art's cathartic power is needed more urgently than ever. Show us the world vanishing on the page, and we may more clearly see sustainable paths ahead. Release in us the pleasure of giving up, and we may find new strength to struggle on.

NOTES

1   Alan Warner, "The Road to Hell," review of *The Road*, by Cormac McCarthy, *Guardian*, November 4, 2006, https://www.theguardian.com/books/2006/nov/04/featuresreviews.guardianreview4.

2   David Moscrop, "The Cold Comfort of Apocalyptic Fiction," *Globe and Mail*, September 23, 2022, https://www.theglobeandmail.com/arts/books/article-a-sense-of-the-apocalypse-filled-our-trip-to-project-mose-in-venice/.

3   Rumaan Alam, *Leave the World Behind* (HarperCollins, 2020), 120–21.

4   Alam, *Leave the World Behind*, 127.

5   Alam, *Leave the World Behind*, 235.

6   Alam, *Leave the World Behind*, 235.

7   Alam, *Leave the World Behind*, 238.

8   Adam Phillips, "On Giving Up," *London Review of Books*, January 6, 2022, https://www.lrb.co.uk/the-paper/v44/n01/adam-phillips/on-giving-up.

9   Doug Henwood, "Dystopia Is for Losers," foreword to *Catastrophism: The Apocalyptic Politics of Collapse and Rebirth*, by Sasha Lilley, David McNally, Eddie Yuen and James Davis (PM Press, 2012), ix–xv.

10 The essay's famous opening line reads: "There is but one truly serious philosophical problem, and that is suicide." Albert Camus, *The Myth of Sisyphus*, in *The Myth of Sisyphus and Other Essays*, trans. Justin O'Brien (1955; Vintage, 1991), 1–138.

11 Andreas Malm, *How to Blow Up a Pipeline: Learning to Fight in a World on Fire* (Verso, 2021), 150.

12   Waubgeshig Rice, *Moon of the Crusted Snow* (ECW Press, 2018), chap. 22, Libby.

13   Franny Choi, "The World Keeps Ending, and the World Goes On," in *The World Keeps Ending, and the World Goes On* (Ecco, 2023), 2.

14   Paul Lynch, *Prophet Song* (Oneworld, 2023), 304.

15   Karl Ove Knausgaard, "All That Is in Heaven," in *In the Land of the Cyclops: Essays*, trans. Martin Aiken, Ingvild Burkey and Damion Searls (Knopf, 2021), 4.

16   "Catharsis, n.," *Oxford English Dictionary* (Oxford University Press, March 2023).

GOOGLE ALERTS

What strikes me as odd is not that everything is falling apart, but that so much continues to be there.

— Paul Auster, *In the Country of Last Things*

From summer 2019 until summer 2022, I used a daily Google Alert to gather online appearances of the terms "democratic crisis," "liberal democracy + crisis" and "crisis of democracy." I set up the alert to track more methodically what seemed to me to be a growing consensus in public debate that democracy was on the verge of collapse. I understood the impulse to declare that democracy was dying. Trump had bullied and blundered through nearly three years in office. (He would be impeached before the end of the year.) The deadline for Brexit was nearing with no implementation plan in sight. Yet I was also skeptical about the precision of the crisis narrative. Was government by the people truly in a do-or-die moment?

When I ended the Google Alerts in 2022, my democracy-in-crisis archive housed thousands of newspaper articles, podcasts, lecture notices and interviews. A long essay on threats to global democracy appearing in the *Straits Times* of Singapore.[1] A syllabus for a Stanford University history class titled "Democracy in Crisis." A Canadian journalist explaining that newsroom closures are "a symptom" of the crisis of democracy.[2] The archive includes reviews of no fewer than twenty books with titles such as *Democracy in Crisis: Lessons from Ancient Athens*; *Twilight of Democracy: The Seductive Lure of Authoritarianism*; *Crises of Democracy*; and what is surely the most-quoted book of the bunch: *How Democracies Die*.[3]

I wasn't surprised that my Google Alerts pointed to op-eds

about civic corrosion in major dailies. Although I approach the archive as a form of political zeitgeist – one articulation of democratic culture in this moment – I assumed the dataset would skew heavily toward elite perspectives. But I was surprised to discover talk of democratic crisis in mass publications targeted at a general readership: *Teen Vogue*, for instance, as well as *Rolling Stone*, *Ms. Magazine*, *Cosmopolitan* and lists of must-watch television programs. Alongside coverage by corporate giants, the archive includes: coverage by oppositional, independent news organizations (*Democracy Now!*, *Jacobin* and the South African *Daily Maverick*, to name a few); Catholic periodicals comparing the crisis of democracy to the crisis of the church; a *Publishers Weekly* profile of the founder of the crime fiction series Low Down Dirty Vote; and a *Sydney Morning Herald* interview with the artistic director of the Sydney Theatre Company who explains why Shakespeare's *Julius Caesar* is the perfect production for our time: "The play opens with democracy in crisis. The democratically elected leader wants to be made emperor for life."[4] The archive is a window onto contemporary political culture.

As the archive grew, I began noticing inconsistencies in my interpretation of its holdings. I saw in the content I was reading and watching compelling reasons for believing that liberal democracy is broken beyond repair, as well as evidence that democracy is *not* in crisis the way so many people say it is. Variability became the central theme of my research: contradictions in my conclusions and in the archive itself. My Google Alerts seemed to turn up a piece, if not a school of thought, for every position along the political spectrum, every temperament, every timescale, every medium and every adjacent issue. How does one organize such a messy collection? Which systems of classification reveal meaningful patterns in the archive?

You could divide material by argument. Over here, for instance, is the tallest stack of common materials, which I've labelled "liberal catastrophism." This is the-end-is-nigh pile. It includes proclamations by political scientists and op-ed columnists declaring that democracy won't last because Trump destroyed essential civic norms, the idea that rising authoritarianism in some places (Hungary, Brazil, Turkey, India) means democracy everywhere is doomed, the suggestion that Britain is in democratic crisis because it left the European Union, that Western democracy is vulnerable to "foreign interference"[5] (namely, Russian agents) and that the rise in voter abstention in France signals a republic in terminal decline. The overall view in this material is liberal in the sense that it equates rule by the people with liberal democracy (i.e., representative government dependent on a sprawling bureaucracy, plus individual rights, plus capitalism). It's catastrophist in its assumption that democracy, if not dead already, is facing imminent collapse.

There are right-wing and progressive liberal catastrophists. The Scottish conservative historian Niall Ferguson asserts that "liberal democracy as we know it won't survive the 21st century" because of worsening economic crises, social media and "the illiberalism of the young" (by which he means student social justice activists).[6] In the *Times of India*, Abhishek Kumar, a student social justice activist, writes that democracy is "dangerously close to crumbling under its own weight" (by which he means right-wing populism).[7] Ironically, considering their characteristic doomsaying, liberal catastrophists are known to suggest that the crisis could be resolved by tinkering with electoral mechanisms. Democracy will have a fighting chance if only we in Canada adopt a proportional representation voting system, if only the US abolishes its Electoral College.

Because liberal catastrophism sets the terms of debate, I apply the broad label of "contrarian" to the stacks of surrounding material. In Canada's *National Post*, the centrist contrarian Michael Ignatieff argues that the brilliance of liberal democracy is its ability to absorb and equalize existential threats.[8] Citizens grumble and quarrel as a matter of existence. Politicians lie from time to time. Partisan rancor is unpleasant and, in exceptional cases, vicious. But partisanship is as old as party democracy itself. Voter participation rates rise and fall cyclically. Liberal democracy strains to keep up with changes in technology, migration and culture. But this is a sign of the system's durability, not an indicator of crisis. In the words of a centrist contrarian op-ed in an Irish daily, given how much worse things could be, it is "both timely and appropriate to generously acknowledge our democratic successes."[9]

Right-wing contrarians depict the crisis of democracy narrative as so much left-wing whining. One *National Review* columnist puts it like this: "For [Obama speech writer] Ben Rhodes to say, in effect, 'OMG you guys, the Electoral College is NOT democratic' is like saying, 'The Electoral College is not a hot fudge sundae.' It wasn't intended to be. The Electoral College, like the Senate, was designed to stop the country from being ruled by a cabal of *Washington Post* and *New York Times* columnists."[10] The *National Post*'s Kelly McParland says the problem isn't the state of democracy, it's that "progressives feel the need to turn everything into a crisis."[11] To the right-wing contrarian, crisis-mongering is political correctness run amok.

The left-wing contrarian agrees that democracy is in crisis; it's just not the crisis diagnosed and feared by liberal catastrophists. An essay by Chris Maisano appearing in *Jacobin* in 2019 argues that "too much of today's 'crisis of democracy' discourse seeks to defend the existing political order against something called norm erosion, as embodied by the eroder-in-chief Donald Trump. But it's this very order that made the emergence of a figure like Trump possible in the first place."[12] Core liberal democratic institutions, he argues, such as anti-majoritarian representative government and corporate control over public resources, create social divisions in which racist, anti-democratic leaders such as Trump can position themselves as fixers.

In several long essays, political scientist Henry Giroux examines "elements of a fascist politics" emerging in the US.[13] Understanding this development requires placing Trump's specific crimes in a context created by Republicans and Democrats (and all Western governments) alike. In the *Daily Maverick* out of South Africa, the sociologist Michelle Williams argues that the favoured solutions of liberal catastrophists – electoral reform, civic literacy campaigns, more critical news media – may be desirable "but are not enough to save democracy."[14] The democratic crisis will continue until fundamental political-economic transformation takes place.

If organizing the archive by argument makes it difficult to detect changes in crisis debates, you could look at the archive as a drama unfolding over time. Act I (summer 2019–spring 2020) is set amid the tumult of Brexit and Trump's first impeachment. The British government's inability to win parliamentary consent for an EU-withdrawal plan is interpreted as an irreparable schism between rulers and ruled. In the *Washington Post*, Ishaan Tharoor explains why "Trump's impeachment battle is part of a bigger global crisis in democracy."[15] Like Netanyahu in Israel, Johnson in the UK and Modi in India, argues Tharoor, Trump does the greatest harm to the body politic by "calling into question the very legitimacy of those opponents challenging his conduct."

Act II (spring 2020–fall 2020) is all about the pandemic. Authoritarian-curious columnists argue that the West's inability to contain Covid-19 is a sign that democracy is ill-equipped to manage the complex problems of the modern world. Conversely, state-skeptical analysts worry about attacks on democratic freedoms in the name of public health.

Act III (summer 2020–early 2021) moves from pre-Trump/Biden election jitters to post-Trump/Biden election jitters. In June 2020, *Rolling Stone* prints a verse from "America," a new blues song with "brutally honest lyrics about the state of American democracy."[16] In the run-up to voting day, prevailing opinion is that the democratic crisis will deepen if Trump is re-elected. Then appears an Election Day piece in *Foreign Affairs* magazine by Stanford University sociologist Larry Diamond proclaiming: "Regardless of who wins the 2020 presidential election, the health of American democracy will not soon recover."[17] As Trump's "Stop the Steal" campaign builds steam, *Salon*'s Chauncey DeVega refers to "Trump's ongoing coup attempt."[18] In the *Week*, Joel Mathis identifies a paradox: "Preserving democracy – as Americans have practiced and understood it – required defeating Donald Trump in the presidential election. But Trump's defeat has only heightened the crisis of democracy."[19] Act III ends abruptly midday on January 6, 2021.

Act IV (early 2021–early 2022) is set against the riot at the US Capitol. It opens with Joe Biden delivering his first State of the Union Address: "The insurrection was an existential crisis – a test of whether our democracy could survive. It did. But the struggle is far from over. The question of whether our democracy will endure is both ancient and urgent."[20] House Representative Abigail Spanberger (a Democrat from Virginia and former Central Intelligence Agent) said, "This is what we see in failing countries . . . This is what leads to a death of democracy."[21] Act IV closes thirteen months later, amid post-mortems of Biden's "dagger at the throat of American democracy" speech marking the first anniversary of January 6.[22] The yearlong drama includes hyperbolic accounts of a so-called armed insurrection, optimistic takes explaining why Trump's grip on politics is slipping and Chicken Little–style forecasts that 2024 will be even worse.

Act V (early 2022–mid 2022) centres on the war in Ukraine, which one scholar calls "a Rorschach test" of people's democratic anxieties.[23] In the London *Telegraph*, Sherelle Jacobs writes that "as Russia smashes international norms," the liberal democratic global order "is coming to a feeble end."[24] Details of the situation in Ukraine get replaced by a more abstract story about the long-standing crisis of war between liberal democracy (led by the US and EU) and authoritarianism (led by Russia and China). Doubtless, I could've written about Act VI (VII, VIII . . . ?) in the months on either side of Trump's 2024 election victory, but I'd shut down my Google Alerts by then.

My archive's five-act drama is a US-centric story. Of course, it's a US-centric world – in terms of media influence and geopolitical heft. That the Google Alerts algorithm produces a surfeit of content about American democracy published by American sources is not surprising. What I didn't expect to find is so much talk about democratic crisis in places other than the United States.

The *Korea Times*, *Al Jazeera*, the *Times of India* and the *Guardian* regularly run big think pieces about the global crisis of democracy. In India's *National Herald*, Prem Anand Mishra compares the democratic crisis in Lebanon to the one in Hong Kong. He talks about "the virulent rise of the far right from Europe to Asia" and says strongmen in office are "blurring the boundaries between liberal democracy and autocratic regimes."[25] You're far more likely to find columnists in the Global South discussing the political crisis in the US within a global, historical context, than seeing it, as Western journalists tend to, as a national struggle between Democrats and Republicans.

There's also a heap of coverage of specific political crises happening within national borders. Democracy in crisis in Poland because of electronic spying. Election rescheduling in Newfoundland creating a provincial "democratic crisis."[26] Crises of democracy declared in Bolivia (coup), Malta (government corruption), India (right-wing populism), Quebec (dysfunctional election system). Kaderi Noagah Bukari argues that the rise of "democratic dictators" changing presidential term limits in Senegal, Cameroon, Uganda and other African countries illustrates "Western majoritarian democracy in crisis."[27] In Israel, Benjamin Netanyahu's fraud trial was repeatedly described as a crisis of democracy (notably, however,

Israel's apartheid system never was). The archive contains reference to democracy in crisis in more than thirty national and subnational units.

You could organize the archive by content type (op-ed, blog post, book review, webinar), by region, by diagnostic versus prescriptive pieces, by tone. (The comedic op-ed with the widest syndication concludes that "the crisis of democracy is the crisis of the restaurant trade and of Heathrow airport. You just can't get the staff."[28]) But whatever organizing principle guides your analysis, what you'll find is striking variation in form and content. There's even a handful of material within the archive defining a different issue in terms of democratic crisis. The eco-crisis *is a crisis of democracy*; poverty *is a crisis of democracy*; racist policing *is a crisis of democracy*; declining enrollments in university humanities programs *is a crisis of democracy*; toxic masculinity *is a crisis of democracy*.

What to make of such expansive variation? The simplest explanation for the many shades and large size of the archive is that democracy *is* in crisis, more or less as the modern Cassandras (future-seeing, but never believed) say it is. David Runciman's book *The Confidence Trap: A History of Democracy in Crisis from World War I to the Present* argues that democracy rarely runs smoothly for long. Major crises nearly toppled liberal democracy on a global scale in 1918, 1933, 1947, 1962, 1974, 1989 and 2008. Then, as now, public intellectuals warned of democracy's imminent end. In the past, governments made structural adjustments in time to avoid irreversible disaster. It remains to be seen whether catastrophe can be avoided this time around. It's tough to ignore "democratic backsliding" with evidence published yearly by Sweden's V-Dem Institute.[29] Does my archive grow in direct relation to the growing likeliness of democracy's end?

No, says Eliah Bures in *Foreign Policy*: "When everything is a crisis, nothing is."[30] Using the same term to describe fundamentally different forces and events confuses more than clarifies. A dip in voter turnout (in France) is not a coup (in Tunisia). Neither is an elected fascist (in India). Yet we conflate all three, along with myriad other trends, and regional and institutional distinctions, when we apply the crisis-of-democracy narrative to them.[31] When economist and podcaster Tyler Cowen was asked if "liberal democratic capitalism is in some kind of crisis," Cowen responded: "Crisis, what does that word mean? There's been a crisis my whole lifetime. People have fallen apart because Brexit happened and Trump was elected. [. . .] They'll be fine. So I've become more optimistic than most people. I feel I've stayed put, they've just gone crazy."[32] On this view, my archive's depth and breadth are signs not of perspicacity but of untrustworthiness.

I reject the view that democracy is on the brink. It's true that the best parts of modern democracy (for example, a social safety net) are eroding, while the worst parts (for example, the power of unelected officials) worsen. But these are shifts within a malleable model that only ever assumes form through social struggle among contending groups. They don't amount to a species-defining break from what official democracy in the West has always been.

Conventional political science suggests that liberal democracy, in addition to requiring representative government, includes features such as an independent press, a pluralist society, a culture of toleration, separation of powers among branches of government and so on. Whether this suggestion is borne of wishful thinking, ideological warfare or sloppy historical analysis is secondary to the fact that defining liberal democracy so expansively is misleading. In fact, in parliamentary liberal democracies like Canada's, the legislative and executive branches of government are fused. For most of the twentieth century, the United States presided over legalized discrimination against African Americans, and the Canadian state refused to provide full rights to Indigenous people – that is, these liberal democracies were anti-pluralist. And in the context of today's media monopolies and well-documented journalistic interference by corporate owners, it's debatable whether "free and independent" is an accurate descriptor of the contemporary press. The character of social rights and a culture of pluralism have differed so significantly during different periods of liberal democratic rule that including them among the model's essential elements obscures more than it reveals.

Shortly after I launched my Google Alerts, I began seeking a more conceptually rigorous, historically grounded way of thinking about liberal democracy's basic architecture. It felt necessary to be able to articulate precisely what was supposedly in crisis before assessing the soundness of the democracy-in-crisis claim. So I asked David McNally, a critical political scientist known for his theoretically rich historical scholarship, to define the strictly essential features of this model of government. McNally's definition isn't the one from your grade ten civics textbook, but you will not find a more precise and illuminating description. In McNally's words: "Liberal democracy means the enshrinement of capitalist property rights within a system of power only formally accountable to elected representatives. In fact, the property rights and privileges of capitalists are legally entrenched and protected by bureaucratic and executive power in ways that hamstring elected officials. Substantively, capitalist property and power are beyond democratic control. So, while the system allows for electoral processes, it locks in structural constraints that prevent those elected from undermining the economic dominance of capitalist interests."[33]

Of course, ordinary citizens appreciate the individual rights and freedoms offered by liberal democracy, protections such as "freedom of speech, freedom of association, freedom from arbitrary arrest," as well as the right to select legislators through periodic elections.[34] But even the liberal theorist C.B. Macpherson highlights the strict limits on popular power within "the real world" of liberal democracy. In Macpherson's words: "The liberal-democratic state is essentially the liberal [i.e., bourgeois] state with a democratic franchise added."[35] Macpherson well understood that "the democratic element in liberal-democracy [. . .] derives less from the necessary requirements of capitalism as a system, and less from the theorists of 'developmental' democracy, than from the *organisation of the working-class*."[36] Social struggle, not built-in structures, determines how democratic liberal democracy will be.

Within the history of liberal democracy, there have been periods when citizenship has been defined narrowly (only men, only whites, only straight people) and more broadly (without gender, racial or sexuality requirements). At times, corporate tax rates were above 50 percent (the late 1960s in the US, the early 1980s in Canada); today they bob and dip around 25 percent. Reproductive rights in the US were not constitutionally protected, then they were, now they're not. The defeat of Roe v. Wade is alarming, gravely harmful and surely a sign of the weakness of the left and the growing power of anti-feminist activism. (In my archive, Katha Pollitt says, "The crisis of abortion is a crisis of democracy."[37]) It is not, however, a repudiation of individual rights as such within liberal democracy, not evidence that liberal democracy as a form of political rule is in crisis.

Shifts in policy direction have resulted in significant differences in people's lives. But they were shifts within a liberal-democratic model. The living conditions of working people and vulnerable groups can get a lot worse (or a lot better) without abandoning the model's essential features. If anything, reproduction of the democratic status quo through nearly two decades of worst-in-a-generation economic recessions, anti-democratic political leaders and a global pandemic is evidence that liberal democracy is emphatically *not* facing imminent, foundational change.

But rather than join the ranks of those who reject crisis-talk for being untrue, I'm interested in the truths people are expressing when they talk about democracy in crisis. An NPR/Ipsos poll published in January 2022 found that "64% of Americans believe U.S. democracy is 'in crisis and at risk of failing.'"[38] A poll by the Morris J. Wosk Centre for Dialogue at Simon Fraser University reports that nearly half of Canadians "say they feel our country is NOT governed democratically."[39] My archive consists of pundits, celebrities, politicians and civil society groups warning that democracy is disappearing before our eyes. Without suggesting that public opinion simply mirrors op-ed pages, I think it's reasonable to assume that a significant proportion of the population, reflecting a range of social groups, shares the same concerns about the fall of democracy that I've been led to by my Google Alerts.

If liberal democracy isn't actually breaking down, why do so many people feel that it is? People tend not to be wrong about their feelings, even if they've misdiagnosed the cause of those feelings. If people are wrong to think that liberal democracy is in a do-or-die situation, might they be right about something else?

My hunch is that "democracy in crisis" has become a clear, concise statement of the much less clear, amorphous feeling that *everything is totally fucked up*. If you spend a week talking with my neighbours, colleagues and students, you will hear someone say: I work hard and still I can't get ahead; politicians don't care what you and I think. Doesn't matter who you vote for, it's the same old crap. Corporations are out to screw us (to say nothing of the banks). You can't trust anyone. And everything on TV sucks. The planet is burning up, the cost of living is skyrocketing and there's nothing you or I can do about it. Things aren't fair. This isn't right.

If we view "crisis of democracy" not as a term aiming at scientific precision but as the ultimate symbol of mass disaffection, or, perhaps, the ultimate code for mass disaffection, it becomes possible to see the truth in its widespread use.

Declarations of democratic crisis obviously include concern about bread-and-butter political problems such as election fraud, lying leaders, unresponsive representatives. (Polling research out of Cambridge University shows dissatisfaction with democracy globally increased to its highest point ever between the mid-1990s and 2020, from 48 percent to 57.5 percent.[40]) My reading of the archive is that talking about a democratic crisis has also become a way of talking about cultural exclusion, economic injustice and fear of the future. When Rev. William Barber II, co-chair of the Poor People's Campaign, appeared on *Democracy Now!* in 2021 to demand trillions of dollars in new spending on social services and fighting climate change, he urged working-class people not to think of their problems in terms of liberal versus conservative. "This is not about that. This is about a moral crisis in our nation, a crisis of civilization, a crisis of our democracy."[41] Saying democracy is in crisis has become a stand-in for lamenting broken promises, unfulfilled potential, lost hope, social fragmentation.

This does not mean that everyone conceives of democratic crisis and resolution in the same manner. The meanings of the words "democracy" and "crisis" are highly contested. It's better to understand all this crisis-talk as shorthand for a myriad of anxieties shaped by where people stand in society. As Stuart Hall put it in a classic lecture on mass media, in the world of political stories and ideas: "There is absolutely no escape from th[e] contestation over meaning."[42] As long as social groups struggle over power they will use language to reflect and advance their interests.

Elites have a long history of proclaiming that democracy is broken because their authority remains partially checked by popular power. For example, in 1975, a Trilateral Commission report titled *Crisis of Democracy*, led by the conservative political scientist Samuel Huntington, built a case for restoring elite control over politics in order to counter social movement gains of the 1960s.[43] In the same ideological vein, in 2016 Andrew Sullivan argued in *New York* magazine that Trump rose to power because America has become "too democratic."[44] Not to be outdone, that same year the *Foreign Policy* columnist James Traub published a piece called "It's Time for the Elites to Rise Up Against the Ignorant Masses."[45] Georgetown University philosopher Jason Brennan's book *Against Democracy* explains why stupid people shouldn't be allowed to vote.

Typically, elites attempt to distract from their principled aversion to democracy by talking about the importance of efficiency, or, as the Trilateral Commission report phrased it, to protect governments from being "overloaded with participants and demands."[46] This Daddy-knows-best mentality informs plenty of articles in my archive suggesting British voters did democracy wrong by choosing not to remain part of the EU. Larry Diamond blames America's "political decay" on "complacent citizens who cannot bestir themselves to vote."[47] Elitism is defended subtly by voices advising that the fix to the crisis of democracy is recommitment to hierarchical institutions (such as the nation, the courts and the market). But as a *Financial Times* op-ed in my archive demonstrates, at times elites can't help from saying the quiet part out loud: "Democracy works better when there is less of it."[48]

By contrast, people in less powerful positions say democracy is broken because we do not see our interests advanced by state and corporate policies or in cultural norms. Popular fear, anger and disaffection are expressed in public opinion research. For example, a Pew Research Center poll released in 2022 showed "at least three-quarters of adults in Japan, France, Italy and Canada say children will be worse off financially than their parents."[49] Hopefulness declines as personal debt balloons.[50] According to an annual Gallup World Poll going back decades, happiness in America is at an all-time low. Forty percent of young adults in Canada are at "a mental health breaking point." Academic studies found "increased loneliness in the general population" through the years of Covid-19.[51]

Yes, elite voices predominate in my Google-generated archive of online material, but the internet is a contradictory field, and my archive includes plenty of working-class and radical views. For example, independent media outlets talk about democracy in crisis when financial pressures threaten their existence. The poet Claudia Rankine says, "It's self-evident that we are in a democratic crisis" because "white supremacy is not only on the rise, but is also empowered."[52] Teachers say the emphasis on standardized testing feeds the crisis of democracy. An anti-apartheid activist says people sense democracy is in crisis because they "do not feel they're getting anything, or very little, from the democracy that we have now."[53] In Canada, Matthew Norris, president of the Urban Native Youth Association, writes that "our democratic institutions are in crisis" because of the divide between Indigenous and settler governments.[54]

When ordinary people say that democracy is in crisis, despite the fact that liberal democracy is reproducing itself effectively, I suspect they're giving voice to the contemporary condition that Dario Gentili calls precarity: economic insecurity and political impotence in the absence of a genuine alternative. Gentili argues that as more people experience deeper precarity without a mass political vehicle promising real change, we find ourselves in an "age of crisis," a period not only of constant tumult, but in which crisis management has become the foremost "art of government."[55]

In fall 2022, I picketed in solidarity with striking education workers in Ontario. As we circled our local Member of Provincial Parliament's office in the November sun, several strikers said words to the effect that the government's use of the so-called "notwithstanding clause" to impose a contract on them and criminalize their strike was "a perfect example of how democracy is in crisis right now." The thing is, the notwithstanding clause is section thirty-three of the very same Canadian Charter of Rights and Freedoms that workers claim was being violated. Using it as the Government of Ontario did was innovative and disgraceful. But it was, by definition, perfectly consistent with the letter of the law in Canadian democracy. Still, I know what the strikers meant.

The ideal of democracy holds out the promise that the people (not elites) know best. Our only hope of achieving freedom and justice for all is by governing ourselves collectively through ongoing processes of experimentation, learning and transformation. In the words of political theorist Anthony Arblaster: "In the twenty-first century, democracy represents political virtue."[56] If the reigning view is that democracy is failing, I take it as a sign of the wisdom of the crowd. As I've said, I'm skeptical that it is failing in the way liberal catastrophists worry about. Yes, every model of politics is vulnerable to the contingencies of history; but at present, I see no alternative political-economic system or movement (from left or right) with the social weight necessary to topple liberal democracy and assume control after the fall. Nevertheless, liberal democracy is failing, as it always has, to greater and lesser degrees, to fulfill its promises of universal freedom and justice.

Saying democracy is in crisis says something (however partially formed) about the potential that *living together* could, and should, be better. It suggests that the idea of democracy, whatever people's ambivalence about it, whatever their misunderstandings and flawed assumptions, is still held in high regard. This isn't a defence of apathy, though I understand why so many people refuse to participate in anything they deem political. But the lament for democracy in crisis, even the dismissive *plague-on-both-your-houses* kind, suggests that people still attribute dignity to democracy, even if they don't believe our own is dignified (or durable). If I'm right that people are using the language of democratic crisis to express a multitude of grievances that go far beyond strictly political ones, I'm encouraged. There is radical potential for change (as well as danger) in assuming that liberal democracy is broken.

NOTES

1   Chan Heng Chee, "World in Transition – The 4 Big Challenges," *Straits Times*, July 4, 2020, https://lkyspp.nus.edu.sg/docs/default-source/ips /st_world-in-transition-the-4-big-challenges_040720.pdf.

2   Em Cooper, "Why the Media Crisis Is More than It Seems. Hear Insights from Tyee Senior Editor Paul Willcocks," *Tyee*, August 7, 2020, https://thetyee.ca/Tyeenews/2020/08/07/Three-Things-Paul -Preview/.

3   Josine Blok, review of *Democracy in Crisis: Lessons from Ancient Athens*, by Jeff Miller, *Polis: The Journal for Ancient Greek and Roman Political Thought* 40, no. 1 (2023): 159–64; Sheri Berman, "The Everyday Decisions that Undermine Democracy," review of *Twilight of Democracy: The Seductive Lure of Authoritarianism*, by Anne Applebaum, *Washington Post*, July 24, 2020, https://www.washingtonpost.com/outlook/the-everyday -decisions-that-undermine-democracy/2020/07/23/691fb72c-c078 -11ea-b178-bb7b05b94af1_story.html; Jan-Werner Muller, "One Damn Thing After Another: The Long Roots of Liberal Democracy's Crisis," reviews of *Crises of Democracy*, by Adam Przeworski, and *Democracy and Dictatorship in Europe*, by Sheri Berman, *Nation*, May 5, 2020, https://www.thenation.com/article/culture/sheri-berman -adam-przeworski-democracy-dictatorship-crisis-book-review/; Adam Tooze, "Democracy and Its Discontents," review of *The People vs. Democracy: Why Our Freedom Is in Danger and How to Save It*, by Yascha Mounk, *How Democracies Die*, by Steven Levitsky and Daniel Ziblatt, *The Road to Unfreedom: Russia, Europe, America*, by Timothy Snyder, and *How Democracy Ends*, by David Runciman, *New York Review of Books*, June 6, 2019, https://www.nybooks.com/articles /2019/06/06/democracy-and-its-discontents/.

4   Massimo Faggioli, "Catholic Synodality as a Response to the Crisis of Democracy," *Catholic Outlook*, November 17, 2019, https:// catholicoutlook.org/catholic-synodality-as-a-response-to-the-crisis-of -democracy/; Alan Scherstuhl, "Trouble Is Her Business," *Publishers Weekly*, August 5, 2022, https://www.publishersweekly.com/pw /by-topic/authors/pw-select/article/90027-trouble-is-her-business

.html; Linda Morris, "Sydney Theatre Company Back on Stage After $10 Million Box Office Hit," *Sydney Morning Herald*, September 30, 2021, https://www.smh.com.au/culture/theatre/sydney-theatre-company-back-on-stage-post-10-million-box-office-hit-20210928-p58vk6.html.

5   Jack Montgomery, "Hillary Alleges 'Russian Involvement, If Not Influence or Interference, in Brexit,'" *Breitbart*, May 6, 2021, https://www.breitbart.com/europe/2021/05/06/hillary-alleges-russian-involvement-if-not-influence-or-interference-in-brexit/.

6   Niall Ferguson, "Munk Debates – Niall Ferguson: Liberal Democracy Will Not Survive the 21st Century," *National Post*, January 9, 2020, https://nationalpost.com/opinion/munk-debates-niall-ferguson-liberal-democracy-will-not-survive-the-21st-century.

7   Abhishek Kumar, "Capitol Shame," *Times of India*, January 11, 2021, https://timesofindia.indiatimes.com/blogs/the-argument/capitol-shame/.

8   Michael Ignatieff, "Ignore the Naysayers. Liberal Democracy Is Built to Last," *National Post*, January 9, 2020, https://nationalpost.com/opinion/munk-debates-michael-ignatieff-ignore-the-naysayers-liberal-democracy-is-built-to-last.

9   Anthony O'Halloran, "Quit the Negative Vibes, and Thank Our Blessings," *EchoLive.ie*, December 13, 2019, https://www.echolive.ie/corkviews/arid-40131320.html.

10 Kyle Smith, "Our Democracy Is Not in Crisis," *National Review*, July 23, 2019, https://www.nationalreview.com/2019/07/american-democracy-is-not-in-crisis/.

11 Kelly McParland, "From Climate Change to COVID, Progressives Feel the Need to Turn Everything into a Crisis," *National Post*, December 21, 2021, https://nationalpost.com/opinion/kelly-mcparland-from-climate-change-to-covid-progressives-feel-the-need-to-turn-everything-into-a-crisis.

12 Chris Maisano, "Bernie's Political Revolution Requires Radical Democratic Reform," *Jacobin*, August 27, 2019, https://jacobin.com/2019/08/bernie-sanders-political-revolution-democratic-reform-electoral-college.

13 Henry A. Giroux, "Impeachment and the Politics of Organized Forgetting: This Attack on Trump Isn't Nearly Enough," *Salon*, January 19, 2021, https://www.salon.com/2020/01/19/impeachment-and-the-politics-of-organized-forgetting-this-attack-on-trump-isnt-nearly-enough/; see also Giroux, "America's Nazi Problem and the End of Policing," *CounterPunch*, June 4, 2021, https://www.counterpunch.org/2021/06/04/americas-nazi-problem-and-the-end-of-policing/.

14   Michelle Williams, "The Crisis of Democracy: The Importance of Reclaiming Democracy from Neoliberal Capitalism and Creeping Authoritarianism," *Daily Maverick*, November 8, 2021, https://www.dailymaverick.co.za/article/2021-11-08-the-crisis-of-democracy-the-importance-of-reclaiming-democracy-from-neoliberal-capitalism-and-creeping-authoritarianism/.

15   Ishaan Tharoor, "Trump's Impeachment Battle Is Part of a Bigger Global Crisis in Democracy," *Washington Post*, October 4, 2019, https://www.washingtonpost.com/world/2019/10/04/trumps-impeachment-battle-is-part-bigger-global-crisis-democracy/.

16   Jon Blistein, "G.E. Smith, LeRoy Bell Capture Democracy in Crisis on New Song 'America,'" *Rolling Stone*, June 26, 2020, https://www.rollingstone.com/music/music-news/g-e-smith-leroy-bell-new-song-america-stony-hill-1020637/.

17   Larry Diamond, "A New Administration Won't Heal American Democracy," *Foreign Affairs*, November 5, 2020, https://www.foreignaffairs.com/articles/2020-11-05/new-administration-wont-heal-american-democracy.

18   Chauncey DeVega, "Nothing to See Here: Why Media Keeps Downplaying Trump's Coup Attempt," *Salon*, December 23, 2020, https://www.salon.com/2020/12/23/nothing-to-see-here-why-media-keeps-downplaying-trumps-coup-attempt/.

19   Joel Mathis, "America's Political Crisis Is Far From Over," *Week*, November 16, 2020, https://theweek.com/articles/950095/americas-political-crisis-far-from-over.

20   Joe Biden, "Remarks by President Biden in Address to a Joint Session of Congress," *The White House*, April 29, 2021, https://bidenwhitehouse.archives.gov/briefing-room/speeches-remarks/2021/04/29/remarks-by-president-biden-in-address-to-a-joint-session-of-congress/.

21   Charles T. Call, "No It's Not a Coup – It's a Failed 'Self-Coup' that Will Undermine US Leadership and Democracy Worldwide," *Brookings*, January 8, 2021, https://www.brookings.edu/blog/order-from-chaos/2021/01/08/no-its-not-a-coup-its-a-failed-self-coup-that-will-undermine-us-leadership-and-democracy-worldwide/.

22   Danielle Kurtzleben, Claudia Grisales and Scott Detrow, "'A Dagger at the Throat of Democracy': President Biden Decries Election Lies," *NPR*, January 6, 2022, https://www.npr.org/2022/01/06/1071063375/a-dagger-at-the-throat-of-democracy-president-biden-decries-election-lies.

23   Liam Kennedy, "Ukraine: A Divided America Seeks Moral Clarity in a War Against Democracy," *Conversation*, April 22, 2022, https://theconversation.com/ukraine-a-divided-america-seeks-moral-clarity-in-a-war-against-democracy-181806.

24 Sherelle Jacobs, "The Ukraine Crisis Is the Final Nail in the Coffin of the Western Liberal Order," *Telegraph*, February 21, 2022, https://www.telegraph.co.uk/politics/2022/02/21/ukraine-crisis-final-nail-coffin-western-liberal-order/.

25 Prem Anand Mishra, "Is It the Age of Protest?," *National Herald* (India), December 2, 2019, https://www.nationalheraldindia.com/opinion/is-it-the-age-of-protest.

26 Greg Mercer, "Newfoundland Election Creates New Democratic Crisis," *Globe and Mail*, March 30, 2021, https://www.theglobeandmail.com/canada/article-newfoundland-election-creates-new-democratic-crisis/.

27 Kaderi Noagah Bukari, "Ghana's Style of Democracy Has Recently Shown Cracks. Here's How to Fix It," *Conversation*, July 18, 2021, https://theconversation.com/ghanas-style-of-democracy-has-recently-shown-cracks-heres-how-to-fix-it-164439.

28 Janan Ganesh, "Dire Tory Leadership Race Shows Western Democracy Just Cannot Get the Staff," *Irish Times*, July 19, 2022, https://www.irishtimes.com/world/uk/2022/07/19/janan-ganesh-western-democracy-has-a-personnel-problem/.

29 D. Erasmus, "Do Democracy: Remake the Wisconsin Idea," *Daily Kos*, April 17, 2022, https://www.dailykos.com/stories/2022/4/17/2092484/-Do-Democracy-Remake-the-Wisconsin-Idea.

30 Eliah Bures, "When Everything Is a Crisis, Nothing Is," *Foreign Policy*, August 1, 2020, https://foreignpolicy.com/2020/08/01/when-everything-is-a-crisis-nothing-is/.

31 Serge Schmemann makes this argument nicely in "Democracy Is Not Facing a Global Extinction Event," *New York Times*, January 11, 2025, https://www.nytimes.com/2025/01/11/opinion/editorials/liberal-democracy-far-right-authoritarianism-populism-europe.html.

32 Damir Marusic and Aaron Sibarium, "The World According to Tyler Cowen," *American Interest*, January 15, 2020, https://www.the-american-interest.com/2020/01/15/the-world-according-to-tyler-cowen/.

33 David McNally, email to author, January 27, 2020.

34 Leo Panitch, "Liberal Democracy and Socialist Democracy: The Antinomies of C.B. Macpherson," in *Socialist Register* 18, eds. Ralph Miliband and John Saville (1981): 163.

35 C.B. Macpherson, *The Real World of Democracy* (House of Anansi, 1965), 64.

36 Panitch, "Liberal Democracy," 153.

37 Katha Pollitt, "6 Ways to Fight for Abortion Rights After Roe," *Nation*, May 30, 2022, https://www.thenation.com/article/society/help-after-roe/.

38   Joel Rose and Liz Baker, "6 in 10 Americans Say U.S. Democracy Is in Crisis as the 'Big Lie' Takes Root," *NPR*, January 3, 2022, https://www.npr.org/2022/01/03/1069764164/american-democracy-poll-jan-6.

39   Steve Anderson, "Canadians Should Pay Attention to the Democracy Crisis in America," *Canada's National Observer*, September 2, 2022, https://www.nationalobserver.com/2022/09/02/opinion/canadians-should-pay-attention-democratic-crisis-america.

40   R.S. Foa et al., *The Global Satisfaction with Democracy Report 2020* (Centre for the Future of Democracy, 2020).

41   William Barber II, interview by Amy Goodman, "'A Moral Crisis': Reverend William Barber on Why Congress Must Pass $3.5 Trillion Bill," *Democracy Now*, September 30, 2021, https://www.democracynow.org/2021/9/30/rev_william_barber_congress.

42   Stuart Hall, "Stuart Hall: Representation and the Media," produced by Sut Jhally, Media Education Foundation transcript (1997): 18, https://www.mediaed.org/transcripts/Stuart-Hall-Representation-and-the-Media-Transcript.pdf.

43   Michel J. Crozier, Samuel P. Huntington and Joji Watanuki, *The Crisis of Democracy: Report on the Governability of Democracies to the Trilateral Commission* (New York University Press, 1975), https://archive.org/details/TheCrisisOfDemocracy-TrilateralCommission-1975.

44   Andrew Sullivan, "Democracies End When They Are Too Democratic," *New York* magazine, May 1, 2016, https://nymag.com/intelligencer/2016/04/america-tyranny-donald-trump.html.

45   James Traub, "It's Time for the Elites to Rise Up Against the Ignorant Masses," *Foreign Policy*, June 28, 2016, https://foreignpolicy.com/2016/06/28/its-time-for-the-elites-to-rise-up-against-ignorant-masses-trump-2016-brexit/.

46   Crozier, Huntington and Watanuki, *Crisis of Democracy*, 12.

47   Larry Diamond, "The Global Crisis of Democracy," *Wall Street Journal*, May 17, 2019, https://www.wsj.com/articles/the-global-crisis-of-democracy-11558105463.

48   Janan Ganesh, "Democracy Works Better When There Is Less of It," *Financial Times*, September 10, 2020, https://www.ft.com/content/f68c13a4-1130-49d5-b3c6-2270711d819e.

49   Laura Clancy, Roseline Gray and Bao Vu, "Large Shares in Many Countries Are Pessimistic About the Next Generation's Financial Future," *Pew Research Center*, August 11, 2022, https://www.pewresearch.org/fact-tank/2022/08/11/large-shares-in-many-countries-are-pessimistic-about-the-next-generations-financial-future/.

50   Statistics Canada, "Hopefulness Is Declining Across Canada: Having Children or Strong Ties to a Local Community Associated with

a More Hopeful Outlook," *Daily*, July 17, 2022, https://www150
.statcan.gc.ca/n1/daily-quotidien/220517/dq220517d-eng.htm.

51 Harry Enten, "American Happiness Hits Record Lows," *CNN*, February 2, 2022, https://www.cnn.com/2022/02/02/politics/unhappiness
-americans-gallup-analysis/index.html; Meredith Bond, "40 Per Cent
of Canadian Young Adults at a 'Mental Health Breaking Point': Poll,"
CityNews, March 14, 2022, https://montreal.citynews.ca/2022/03/14
/canada-mental-health-young-adults-covid19-poll/; Roger Patulny
and Marlee Bower, "Beware the 'Loneliness Gap'? Examining Emerging Inequalities and Long-Term Risks of Loneliness and Isolation
Emerging from COVID-19," *Australian Journal of Social Issues* 57, no.
3 (2022): 562–83.

52 Andrew Solomon, "Claudia Rankine Tries to Answer the Impossible,"
Interview, February 24, 2022, https://www.interviewmagazine.com
/culture/claudia-rankine-tries-to-answer-the-impossible.

53 Raymond Suttner, "Democracy Is in Crisis – Moving Beyond 'Disillusionment' Means Recreating Our Political Life," *Daily Maverick*, August 9, 2022, https://www.dailymaverick.co.za/article/2022
-08-09-democracy-in-crisis-recreate-political-life-to-move-beyond
-disillusionment/.

54 Matthew Norris, "Blockades Aren't the Crisis. It's the Crumbling Legitimacy of Canada's Democracy," *Tyee*, February 28, 2020, https://
thetyee.ca/Opinion/2020/02/28/Blockades-Arent-The-Crisis/.

55 Dario Gentili, *The Age of Precarity: Endless Crisis as an Art of Government*, trans. Stefania Porcelli, in collaboration with Clara Pope (Verso,
2021), xvii.

56 Anthony Arblaster, *Democracy*, 3rd ed. (Open University Press, 2002),
10–11.

MY STRUGGLE AND *MY STRUGGLE*

> For the heart, life is simple: it beats for as long as it can. Then it stops.
> – Karl Ove Knausgaard, *My Struggle: Book 1*

I might have had a midlife crisis without realizing it at the time. In my late-thirties, over a span of two years, I pledged enduring love to three different women. I moved houses and cities multiple times, fell off the wagon and prepared to quit my job so I could move to the other side of the world. I thought I was doing just fine.

According to Elliott Jaques, the Canadian psychoanalyst who coined the term "mid-life crisis" in 1957, we're prone to "rapid transition" in our late thirties because, for the first time, we face "the reality and inevitability of one's own eventual personal death."[1] Even young children understand death at an abstract level. But, long into adulthood, most people repress the enormity of the fact that they will die.

By our mid-thirties, the reality of death is no longer easily avoided. Our parents' health declines; they die. Our bodies show the first signs of irreversible decay. The early thrill of marriage and parenthood fades. Careers grow stale. Big plans wither. It's a decisive moment – a crisis – in the life course.

How to respond to this new world, in which death – both real and symbolic – looms large? Fall to pieces, seize up, kill yourself? "Breakdown may be avoided," suggests Jaques, "by means of strengthening manic defences."[2] Buying a sports car, leaving your wife for your secretary – classic midlife crisis stuff. There's also the possibility of resolving the crisis by integrating mortality's dark truth in a fuller life. Jaques points to Dickens and Shakespeare turning to more "tragic and philosophical content" later in their careers, as compared to the "lyrical and descriptive content" of their youth.[3]

The autumn in which I was thirty-seven, death wasn't much on my mind. I didn't feel old or think much about aging. I had all my hair. I was unmarried and childless. I was on my first sabbatical from teaching, living in New York City and writing a book. My partner was making a film. Each morning, I biked past brownstones and junkyards en route to a shared writers' space.

In November, I fell in love with a writer. In January, I left my partner. As I packed books into boxes in our sublet, she asked if I'd lost my mind. I listened to her words but trusted my reason. I was following my destiny.

My new relationship lasted five months. One sunny morning, after playing tennis, I told the writer I'd fallen in love with that I didn't love her anymore. I still loved my ex and wanted to try again with her.

My ex didn't reject the idea but couldn't start in earnest before moving past the pain I'd caused. So began a year of being together but not being together.

Alone in my apartment back in Canada that fall, I sat on a velour easy chair and read the first five volumes of Karl Ove Knausgaard's 3,500-page autobiographical novel, *My Struggle*. I'd long avoided the book, repulsed by the author's apparent self-absorption and titular reference to Hitler's memoir-manifesto. But while driving through the Adirondacks to visit my ex (or not-ex?), I heard Knausgaard interviewed on the radio and became intrigued by his project. Before finishing ten pages, I was obsessed with his story and his way of telling it.

That fall, I read Knausgaard every night after jogging. Most days, on campus, I read pages between student meetings. I read Knausgaard while watching Trump's first election win. I read Knausgaard while refusing to consider the possibility that I might have feelings for a graduate student. *My Struggle* was the most stable part of my life.

Was I going through a midlife crisis? You laugh. But I'm hesitant to apply the label to my experience because I didn't think about it that way at the time. Crises involve real-world events *and* interpretations of those events. Situations on their own, apart from how we define them, are simply that: situations. I saw plainly the dramatic swings in my life; I just didn't view them as adding up to a crisis.

My most rational, fully conscious thoughts during that period didn't explore common midlife stressors. Not once in my internal dialogues about my disruptive behaviour did I think in terms of life transitions or milestones. Not in chats with friends, not in therapy sessions. Today, I shake my head while recalling the self-narrative I performed at that time because so much of it consisted of lies.

I lied about my affairs. Lied about my drinking. Lied about my confidence in this or that decision. I told truths, too. But truth was mixed with fears and wishes, and the compulsion to manage others' impressions of me.

Was I ever fully honest and consistent while moving through years of deceit and turmoil?

My devotion to Knausgaard was pure. I spent months sitting with *My Struggle*, looking forward to getting back to it, and reflecting on its pages. What so captivated me?

In the *New Yorker*, James Wood says *My Struggle* is a book about death. Obviously, the death of the author's father drives the narrative. However, Wood argues, "the more pervasive struggle is with death itself, in which writing is both weapon and battlefield. Writing promises to rescue moments from the march of time, but serious writing also lays bare, examines, dramatizes – and, in this sense, seems to prolong – that death journey."[4]

Poet and novelist Ben Lerner says Knausgaard writes as though "everything seems equally worthy of differentiation" – memories of comic books, sunsets and forest smells; the virtue of soft versus crunchy cornflakes; the feeling of falling in love; hours of cleaning up after a dead man; masturbating; writing; not writing; making tea.[5] If mortality is *My Struggle*'s main character, the book's glacial pace and obsessive attention to detail are acts of rebellion against death.

Only recently, at forty-two, while thinking about what our literary obsessions reveal, did I see in those tumultuous years my preoccupation with death. If you'd asked me at the time, I would've told you I didn't think about my youth ending, my going "over the hill," my dying. My obsession with *My Struggle* tells a different story.

Reconceiving of those years in terms of midlife crisis doesn't absolve me of responsibility for the harm I caused then (nor am I suggesting that everything I did is regrettable). Rather, the lens of midlife crisis reveals logic (however flawed) in a period of my life that once seemed largely incomprehensible. I wouldn't have grasped that logic, or learned from it, if I hadn't reflected on why I was so enchanted by Knausgaard's book.

You might want to ask yourself about your searing passion for the Toronto Raptors. Or reflect on what you get, besides flowers, from obsessive gardening. Or maybe it doesn't work that way. Maybe the secrets of our fixations are recoverable only in retrospect. In any case, the complexity of lives, in crisis or not, is best understood when we examine indirect expressions of thought and feeling. In the words of artist Amy Krouse Rosenthal, who died at fifty-one: "Pay attention to what you pay attention to."[6]

NOTES

1   Elliott Jaques, "Death and the Mid-Life Crisis," *International Journal of Psychoanalysis* 46, no. 4 (1965): 502, 506.

2   Jaques, "Death and the Mid-Life Crisis," 511.

3   Jaques, "Death and the Mid-Life Crisis," 504.

4   James Wood, "Total Recall," review of *My Struggle*, by Karl Ove Knausgaard, *New Yorker*, August 6, 2012, https://www.newyorker.com/magazine/2012/08/13/total-recall.

5   Ben Lerner, "Each Cornflake," review of *My Struggle*, vol. 3, *Boyhood Island*, by Karl Ove Knausgaard, trans. Don Bartlett, *London Review of Books* 36, no. 10 (May 2014): https://www.lrb.co.uk/the-paper/v36/n10/ben-lerner/each-cornflake.

6   Amy Krouse Rosenthal (@missamykr), "for anyone trying to discern what to do w/ their life: PAY ATTENTION TO WHAT YOU PAY ATTENTION TO. that's pretty much all the info u need.," Twitter, March 15, 2013, https://twitter.com/missamykr/status/312564535242395648?lang=en.

AGAINST FATALISM

In a burst of creativity in autumn 1962, Sylvia Plath wrote most of the brilliant poems that were published posthumously in *Ariel*. She died by suicide on February 11, 1963. Plath's journal entries and letters from the final months of her life suggest that she recognized herself to be writing from crisis.

One week before killing herself, Plath wrote to her friend Marcia B. Stern of the dire situation Londoners were in because of that winter's extreme cold: "Everything has blown & bubbled & warped & split – accentuated by the light & heat suddenly going off for hours at unannounced intervals, frozen pipes, people getting drinking water in buckets & such stuff – that *I am in a limbo between the old world & the very uncertain & rather grim new*."[1] The previous fall, Plath quoted from her poem "Elm" in a letter to her mentor, Olive Higgins Prouty: "I know the bottom, she says. I know it with my great tap root." But, in October 1962, Plath described herself emerging intact and energized from the crisis of her broken marriage: "I shall forge my writing out of these difficult experiences – to have known the bottom, whether mental or emotional, is a great trial, but also a great gift."[2] She resolved to persevere.

In winter 1963, Plath saw precisely where she stood, that is, on the line separating life from death. One week before dying, she wrote her final poem, "Edge," which many have since read as a suicide note.[3] The poem begins: "The woman is perfected. / Her dead / Body wears the smile of accomplishment."[4] Yet, in that same period, Plath wrote to her mother: "I am a genius of a writer, I have it in me. I am writing the best poems of my life; they will make my name."[5] While she was enraged and humiliated by her husband's infidelity, aching from the cold, exhausted by caring for two young children, Plath celebrated her sense of making an artistic break-through, the reward of her persevering through the storm.

There is something profoundly moving about Plath's awareness of teetering on the edge and not only seeing her situation clearly, but rendering it into art. In her biography of Plath, Janet Malcolm writes that most of us, at some point, have been "crazed" by love lost, sexual jealousy, plunging despair: "But what few of us have ex-perienced during the progress of the illness is a surge of creativity, empowering us to do work that surpasses everything we have done before, work that seems to be doing itself."[6] At the same time as Plath described death's irresistible attraction, her writing made her feel newly alive, or at least capable for the first time of achieving the literary heights to which she'd devoted her life.

This can sound like cheers from the Plath death cult. I swoon at Plath's dark final messages, picture her saying goodbye. And I do. The thick masking tape around the door frame. Bread and milk left on the floor for her children. Preparing a soft place to lay her head in the oven. Such care, consideration, precision in the act. ("Suicides have a special language," wrote Plath's friend and fellow confessional poet, Anne Sexton: "Like carpenters they want to know *which tools*. / They never ask *why build*."[7])

But it's not Plath's death, at least not exclusively her death, that makes her biography relevant to my interests in this book. Rather, it's that when she was writing, right up to February 11, she hadn't done it, wasn't dead, was still, at least partially, committed to life. She was living through a crisis in the original Greek sense of the term. Her situation had become so severely unstable, so unmanage-able and dangerous, that she'd arrived at a transformative moment: the decisive point between recovery and death.

If the concept of crisis is to mean something qualitatively dif-ferent than disaster, catastrophe and other linguistic markers of wreckage, *badness*, it will be because crisis implies the open-end-edness of intensely disruptive, unsustainably threatening periods of uncertainty. That the patient dies is not the crisis. The crisis is the life-or-death struggle, the period in which the illness could still resolve in vastly different endings.

Plath could have lived, either by being rescued from the domestic cooking gas that killed her (as some have suggested she'd wanted to be) or by not committing suicide in the first place. Why is this truism ignored in the reigning interpretation of Plath's life, death and art? The popular story of Plath says, instead, that her crisis *could not* have ended differently but was leading unavoidably to her suicide, that Plath was fated to die young.

In the "doom myth" of Sylvia Plath, as Melissa Bradshaw calls it, Plath's art was inseparable from her personal thoughts.[8] As the poetry became darker, more focused on death, Plath announced her surrender to the void. The scholar Emily Van Duyne suggests that there are two variations of the doom myth: "Plath is either dead and writing her poems from beyond the grave, or else her poems kill her in the culmination of her life as a stage set, a work of high art."[9] In Heather Clark's exhaustive recent biography of Plath, she summarizes previous biographies as focusing "on the trajectory of Plath's suicide, as if her every act, from childhood on, was predetermined to bring her closer to a fate she deserved for flying too close to the sun."[10]

The doom myth relies heavily on a literary sleight of hand performed by Plath's husband, the renowned poet Ted Hughes. After Plath's death, Hughes, as Plath's literary executor, significantly changed Plath's selection and ordering of poems in the first-ever published edition of *Ariel*. Crucially, against Plath's instructions, the Hughes-edited collection concludes with the poems "Edge" and "Words," neither of which Plath indicated should be included in *Ariel* at all, both of which are easily read as Plath's farewell to the world. (The latter poem meditates on how words live on after the poet's death.) With the vocal exception of Ted's sister and a handful of acolytes, it's now commonly understood that Hughes made the changes to hide unfavourable images of him and to portray Plath's death by suicide as being long-planned, even fated.[11] It certainly set the terms for how to read *Ariel* in the decades after it was published.

For example, in a 1965 review of (Hughes's edition) of *Ariel*, the art critic Al Alvarez, a friend of Plath's, wrote that Plath's death was "in a way, inevitable, even justified, like some final unwritten poem."[12] She sacrificed herself to her art. According to the poet Robert Lowell, Plath's former teacher, Plath's fate was evident years before her death. Lowell describes Plath not in human terms, writing that she was "hardly a person at all," but as a tragic figure from Greek mythology, a Dido or Elektra or Medea.[13] Lowell's characterization is partly a nod to Plath's frequent use of mythic imagery in her poetry, and partly to emphasize Plath's seemingly superhuman talent. But, as Elizabeth Roodhouse explains, "once Plath has been embedded within this pre-existing narrative structure, Lowell continues on to imply that Plath's suicide [like Dido's] was inevitable."[14] In Anne Stevenson's biography, published nearly thirty years after Plath's death, a chapter on the poet's youth ends by suggesting that "the idea of suicide formed in her mind like the ultimate and irrevocable fig" (the fig being an image representing unavoidable fate in Plath's novel *The Bell Jar*).[15]

Fictionalized versions of Plath's story invariably splice scenes of her death into scenes spanning her life. Whether in Emma Tennant's widely-praised novel *The Ballad of Sylvia and Ted*, or the widely-panned Hollywood biopic *Sylvia* (starring Gwyneth Paltrow and Daniel Craig), the message hardly needs spelling out: Plath was always going to kill herself, the question being not if, but when. She was "dead *while* she lived."[16]

The doom myth of Plath is understandably appealing. We know how the story ends, so we read causation backwards onto the days leading up to the end, making it all but impossible to imagine a different outcome. Plath's mental illness and the suicide she survived in 1953, Hughes's abusiveness, his recent abandonment of

Plath and their children at the start of his affair with Assia Wevill, the disturbing themes in Plath's poetry – these features of Plath's life become fateful omens of her coming fall. The doom myth is an expression of fatalism, "the idea that what happens (or has happened) in some sense *has* to (or *had* to) happen."[17]

We know from our own lives the feeling that things were fated to be. Each time I've fallen in love, I've felt as though such romance couldn't have been otherwise; that every choice and chance encounter making up my life to that point was leading, as though being guided by an external force, to this connection. The history of literature and film tells me that my love-borne belief in the power of fate is not unique. Of course, when things go badly, when love is lost, dreams go unfulfilled, the job you want goes to someone else, there is consolation in believing that it wasn't meant to be, that the gods, or the stars, have a different plan. *Everything happens for a reason.* Is there a more comforting nugget of folk philosophy? Paul Auster, one of the most celebrated authors of our time, built his career by exploring the human tendency to turn randomness, chance, coincidence, chaos into coherent narratives of destiny.[18] John Jeremiah Sullivan, an American essayist called the voice of his generation, said in a 2024 interview that he thinks about "life and fate in a Spinozistic way: that it's all unfolding, and that there's probably no free will. So it had to be just as it was. For any of it to exist, all of it had to exist."[19]

For centuries philosophers have advised on what to do about the hand dealt by fate. The Stoic says: accept it. To Epictetus, in the first century CE, the core of philosophy is "precisely in learning to desire each thing exactly as it happens."[20] What's virtuous is what's given. The human project is to embrace, to find peace in, or, if not, at least to bear with equanimity the world as we find it. This is the essence of Montaigne's sixteenth-century quietism: the point of life is not to remake the world, but to remake ourselves to welcome the world as it is. (The philosopher Lea Ypi argues that Stoicism's deference to prevailing conditions explains why it appears so often in a certain kind of popular self-help book under late capitalism.)

Machiavelli, who died five years before Montaigne's birth, used violent misogynist imagery in advising rulers that while the Goddess of Fortune can never be fully subdued, she can be partially managed by the prince willing to "cuff and maul her" as necessary.[21] Fate, coincidence, the natural calamities that befall us, these are aspects of human existence that will be with us always. The job of the prince is to put in place ameliorative mechanisms designed to handle fortune's cruel hand. You cannot stop the Goddess of Fortune from lashing your city with weeks of rain. However, when the weather is fine, you can build the levies high in anticipation that floods will come before long. To the psychoanalyst Paul Dubois, an early twentieth-century pioneer of cognitive behavioural therapy, "Patience towards unavoidable events, depending neither upon us nor upon others, is synonymous with *fatalism*; it is a virtue, and it is the only stand to take in the face of the inevitable."[22]

Doubtless there were unavoidable forces in Plath's life. Her brain was wired a certain way. The weather in London was what it was. Her husband had left for another woman. There are givens. And in my own life, I'm constantly speculating on what they add up to, which elements of my existence are open to change, and which have already been determined. I assume that the number of my remaining heartbeats depends somehow or other on the number of sit-ups I do and the number of French fries I avoid eating. But maybe my lifeline is already written, indelibly, on my palm.

Chloe Benjamin's novel *The Immortalists* opens with a fortune teller revealing to four young siblings the date upon which each will die. Even compared with other examples of straightforward fortune-telling in literature (the witches' prophecy in *Macbeth*, say, or John Trause's warning in Auster's *Oracle Night*, "Be careful, Sid. [. . .] Those notebooks are very friendly, but they can also be cruel"[23]), the prophecy delivered by Benjamin's "rishika" is blunter than a sledgehammer to the face. On the one hand, it's the exactitude of the prophecy that makes it easy to dismiss as fantasy, a fun, perhaps, but ultimately fanciful literary game. On the other hand, it's precisely the idea that the future has already been written, not in broad strokes but in fine detail, that chimes with ancient suspicions buried in human consciousness. Haven't you wondered whether the date of your death is out there somewhere chiselled in a cloud? You just haven't yet come across the rishika who can tell you your last day. (Whether you'd want to know the date, and how knowing would change you, are different questions; indeed, the ones Benjamin explores in her book.)

The problem is that too much appreciation for what's given, too much emphasis on the fact that some things are out of our hands, can lead us to defer too much to fate. We begin to see the future as fixed.

The myth of Plath's doom is an ideal case of such misreckoning. In it, the death drive is made to seem as though it alone ruled Plath's movements, to the point that struggle in Plath's life is erased. Yet there was struggle, an attraction to renewal, what one scholar calls "the move toward rebirth" expressed in some of the *Ariel* poems.[24] In the final stanza of "Wintering," for example, the poem that Plath chose to conclude the *Ariel* collection, life emerges victorious. Its quiet promise is beautiful: "Will the hive survive, will the gladiolas / Succeed in banking their fires / To enter another year? / What will they taste of, the Christmas roses? / The bees are flying. They taste the spring."[25] You might object by saying, So what? It's a poem, a work of art, separate from the interpretations and intents of Plath the person. Yet it's precisely the poem that Hughes chose to conclude *Ariel*, "Edge," that he and others pointed to as Plath's commitment to die.

In Elisa Gabbert's words, Plath never stopped being "the joyful hedonist who took such sensual pleasure in living."[26] Two weeks before dying, Plath broke down in a neighbour's apartment, saying,

"I don't want to die. There's so much I want to do."[27] Heather Clark counters the doom myth by arguing that Plath's "near-constant theme is not depression, but the holy quest to become a writer."[28] Yes, the death drive surged in Plath many times during her adulthood; but it was not the only, not even the primary force in Plath's life. Clark guesses that Plath began preparing in earnest to commit suicide around February 8, when she wrote her last letter to Hughes and bought thick tape for sealing off the kitchen. Clark emphasizes, though, that even in the early hours of February 11, the outcome of the plan was uncertain. The future was open, in Van Duyne's words, "with the possible still present, the present still possible."[29] Plath's crisis could have been resolved in a way that satisfied "her hunger for existence."[30]

It's rarely easy to conceive of the future as open-ended. Even if we accept the philosophical distinction that, as opposed to the past, "there is something we can do about the future,"[31] or the folk wisdom that "anything can happen," we live by assuming that we actually have a pretty good idea of what will come next. If not the specifics, at least the outlines of the future seem clear to us based on our knowledge of what has happened in the past, and where present trends point. How else could we live? How could we start relationships, commit to political causes, train for the workplace, if we didn't trust in our ability to foretell the future?

Once the future is realized, Plath dies by suicide – the not-yet-happened becomes first present, then past – it becomes even more challenging to conceive that things might have turned out differently. History has a knack for making itself appear inevitable. Because we know how the Second World War ended, it takes keen mental effort to appreciate the reality of uncertainty prevailing in the early 1940s, when it looked as though the Nazis may well win.

It's one thing to declare that European colonization wasn't an historical necessity, that global history could have unfolded otherwise; but writing from where we stand today, it's not easy to conjure visions of a time when that otherwise still lived.

There's a reason that good counterfactual historical novels are both rare and strangely fascinating. They demand that readers suspend reality twice, both in the standard literary practice required by audiences of all novels, and in the momentary surrender of what we know *really* happened. When done well, for example, in Philip Roth's *The Plot Against America*, in which the isolationist, fascist Charles Lindbergh defeats Franklin D. Roosevelt and becomes president in 1940, the novel succeeds not only by imagining a plausible, compelling alternative story about what politics in America might have been, but by bringing us back to a point in history that truly existed – in Roth's case, America before it entered the war – and helping us conceive of that time, now closed by history, as open to the forces of human agency and chance. Laurent Binet's *Civilizations* does something similar with the (counterfactual) history of colonization, but by imagining a reverse power grab (the Indigenous peoples of the Americas take over Europe), rather than something more radically revisionist, such as imagining world history without imperialism.

Why bother with the mental labour required to stay alert to the fact that the past might have turned out differently, that Plath was not fated to die young? It's not as though we can change what happened years (or even seconds) ago. "There is nothing we can do about the past," writes the philosopher Derek Lam.[32] Whereas the future is defined by openness, the past is defined by "fixity." I'd argue, somewhat paradoxically, that it's by pushing ourselves to see what is now fixed (i.e., the past) as having once been open (i.e., when the future of a specific past was still in the future) that we can see most clearly the open future of our present.

Seeing the future as open-ended is especially important in

times of crisis. Losing sight of the reality of uncertainty makes periods of extreme danger and instability appear as though they are bound to end in ruin. In truth, it's the precariousness of the period that holds out hope that the crisis could end in recovery, rebirth, revolutionary advances in personal or social circumstances. The open, promising side of crisis is erased in fatalistic versions of the past, making it all the harder to identify positive potential in crises of the present.[33]

At the individual level, this misrepresentation of Plath's crisis debases a singular life. It obscures the sexism of the mid-twentieth century and the role of Hughes's abusiveness in shaping Plath's world.[34] It invests mental illness with the power of the Erinyes, Greek goddesses of vengeance.

At the social level, fatalism during times of crisis cultivates passivity by reducing what is, by definition, a two-sided phenomenon – the struggle between recovery and death – to the unidirectional unfolding of either one or the other paths in the struggle. In practice, this might mean ignoring the crisis state of the environment on the assumption that humans will be fine, we'll recover eventually, we always do; or ignoring the nature of the crisis on the assumption that the irreversible eco-apocalypse has already been triggered. Both are expressions of fatalism. There are facts, hunches, reasoned arguments to support both opposing views. The weakness of each alone is its partiality. The truth of the crisis as a whole lies in the combined truths of each partial view.

I hear concessions to fatalism among some critics of the movement for Palestine liberation and decolonial movements more broadly. These critics pose as realists: Look, Israel as a Jewish ethnostate exists, and it's a key ally of America; it isn't going anywhere. Or, Look, it's impractical to advocate decolonization of the settler-colonial states in North America. Millions of people in Canada and the United States aren't just going to "go back to Europe," so get real. Such criticisms of radical visions betray an inability or refusal to see the struggle, the dynamism, that exists in reality. By their logic, the sphere of possibility closes around the current global order, as if our future were fated to such restrictions. What has happened to the human capacity for reason and change? Chance? Surprise?

On the *Past, Present, Future* podcast, the political scientist David Runciman argues that the force of digital technology, big data and AI has irreversibly weakened the human capacity for reason, which stunts imagination and erodes democracy.[35] Runciman says that the tech revolution has made humans less able to draw mental connections and problem solve – "it just has" – and there's no recovering our lost rationality: "I don't think it's coming back."[36] We may come "to love our digital servitude" in ways that would make Aldous Huxley blush,[37] or perhaps the rise of techno-authoritarianism will look more like one of the grimmer *Black Mirror* episodes. Either way, the general direction we're heading is clear. Runciman wouldn't call himself a fatalist, but there's more than a whiff of fatalism in his assumption that a line has been crossed – that human-like rationality invested in machines "has acquired a life of its own" – severely constraining the potential of our species.

Runciman's guest, the philosopher Lea Ypi, picks up on Runciman's fatalism and dispatches his view precisely in these terms:

Fine. It [human rationality] has escaped. Is the escape irreversible or not? Because if you think it's irreversible, okay fine – let's close down the podcast, go home, get drunk and wait for everyone to die. If you say it's reversible, the next question is: What are you doing to reverse it? And that's the question that people need to ask themselves. There's no point in going around, everyone beating each other up saying, "Oh, it's so terrible – it's escaped – it's out of control." These machines are designed by humans [. . .] could they be designed differently? Yes or no. Yes. If they could be designed differently, what are we doing to design them differently? [. . .] We all have individual responsibilities in upholding the kind of social structures that make up our lives. We don't get out of them by just saying, "Oh, it's so terrible, we can't do anything." Because, actually, we can do something. All of us.[38]

Runciman's dim view of the future may be a compelling prediction, but there's a categorical difference between prediction and fatalism. The former is a best guess based on current trends that implicitly acknowledges its fallibility (because it's conceived as a contestable assertion; because it acknowledges that current trends can change; because we've witnessed better and worse predictions in the past; because science advances by making better predictions based on new data and theoretical frameworks). The latter sees the future as inevitable, the realm of certainties already unfolding. Predictions are vital; fatalism is deadening.

I'm reminded of Auster's description of what it's like to read the part of a detective novel before the mystery is solved, as opposed to the parts after the case is closed. When the case is open, "everything could be significant. It makes you very alert as a reader, and that alertness brings pleasure."[39] Solutions in many detective novels "are usually quite boring and uninteresting." There's a letdown when the mystery is solved "because it has to be one thing and not all things. But when you have all things in your mind as a potential answer, that's when you feel most alive." By asserting closure prematurely, declaring the case closed when the mystery still lives, fatalism denies the enlivening experience of seeing everything as potentially significant.

In our troubled times, when catastrophe can feel unavoidable, a politics of progress requires conceiving of crisis as something more complex than catastrophe and destruction. The richest interpretations of contemporary crises will attend to the potential for rebirth contained therein. By rejecting the fatalistic myth of Plath, let's resist the lure of fatalism everywhere. "Hope," writes Rebecca

Solnit, "is an embrace of the unknown and the unknowable, an alternative to the certainty of both optimists and pessimists. Optimists think it will all be fine without our involvement; pessimists adopt the opposite position; both excuse themselves from acting. It's the belief that what we do matters even though how and when it may matter, who and what it may impact, are not things we can know beforehand."[40]

The case of Plath, whose struggle did end in destruction, may seem like a poor model from which to build hope in renewal through crisis. But it's precisely because Plath's future remained open right up until her final breath, long after what's permitted in the doom myth of Plath, that a re-examination of Plath's crisis is conceptually illustrative and politically motivating.

The more I get thinking about the general relevance of Plath's biography to the experience and concept of crisis, the more I find myself reflecting on the political valence of Plath's poetry and the controversy surrounding it. The doom myth of Plath misrepresents her as a poet interested only in her own emotions. In fact, Plath conceived of her poetry as speaking to concerns at multiple registers, personal and social. I've written this book to experiment with inquiry that moves between levels of abstraction more fluidly than what is permitted by my research constrained by the disciplinary conventions of social science. In the explorations of this project, I've found encouragement and caution in Plath's experimental approach to boundary-blurring in the name of expressing the unity of self-and-social.

For example, in Plath's poem "Daddy," she uses Nazi imagery to represent her father and her husband. Plath writes of her father: "I have always been scared of *you*, / With your Luftwaffe, your gobbledygoo. / And your neat moustache / And your Aryan eye, bright blue. / Panzer-man, panzer-man, O You –"[41]

Later in the poem, after the father dies, the narrator-Plath attempts suicide. A stanza later, she reports that "they stuck me together with glue. / And then I knew what to do. / I made a model of you [this "you" is Ted Hughes, the Daddy-cum-husband], / A man in black with a Meinkampf look // And a love of the rack and the screw. / And I said I do, I do."[42] Several times in "Daddy" Plath compares herself to a Jew (though not literally, says Jacqueline Rose, but in the sense of being "one without history or roots"[43]).

In "Lady Lazarus," Plath addresses "Herr Doktor" and "Herr Enemy." Toward the end of the poem, the narrator again assumes a Jewish identity, but now the voice is vengeful and, apparently, almighty: "Herr God, Herr Lucifer / Beware / Beware. // Out of the ash / I rise with my red hair / And I eat men like air."[44]

Critics called Plath's Holocaust metaphors "monstrous, utterly disproportionate." One argued that Plath's references to Nazis were "'empty' and 'histrionic,' 'cheap shots,' 'topical trappings,' 'devices' which 'camouflage' the true personal meaning of the poems in which they appear."[45] Being scolded, cheated on, talked down to, even smacked around is incomparable to the mass extermination of the Jewish people. Plath's mother, Aurelia, wrote to her daughter that the poems were too violent, exaggerated, in bad taste. Responding to her mother, Sylvia Plath wrote: "Don't talk to me about the world needing cheerful stuff! What the person out of Belsen – physical or psychological – wants is nobody saying the birdies still go tweet-tweet but the full knowledge that somebody else has been there & knows the worst, just what it is like. It is much more help for me, for example, to know that people are divorced & go through hell, than to hear about happy marriages. Let the *Ladies Home Journal* blither about those."[46]

It's remarkable that Plath's defence of the poems isn't based on their aesthetic or biographical value, but instead on their political relevance. It's as if to Theodor Adorno's proclamation that "to write poetry after Auschwitz is barbaric,"[47] Plath is saying, I hear you. And you're right that art cannot go on being what it was. But if art creates bonds between those in pain, if it represents the true horror of our existence, it may be indispensable to manage suffering across personal and social spheres. Jacqueline Rose argues that using symbols of patriarchy, fascism and death to slide between autobiography and world history is the point of "Daddy," not a failing or mark against it. If we understand Plath's poetry as experiments in representing "metaphor, fantasy, and identification," then the question is "not whether Plath has the right to represent the Holocaust, but what the presence of the Holocaust in her poetry unleashes, or obliges us to focus, about representation as such."[48]

The confessional poetry movement, of which Plath was a leading light, made inquiries into the human condition by representing experience, especially pain, at various orders of existence simultaneously. For example, Robert Lowell's "intense self-explorations in *Life Studies*" were "a source of metaphors for understanding aspects of the public world."[49] Confessionalism is sometimes dismissed for being egocentric. But, says Steven K. Hoffman, such criticism misses the confessional poet's "immersion in the primary existential conditions of life at any time and also in specifically modern difficulties,"[50] such as postwar collective trauma, Cold War–era nuclear threats and racial conflict. The confessionals use the "implicit interchange between private and public realms" to connect self and other.[51]

Plath's letter to her mother asserts that there's comfort in knowing that others suffer as she suffered. (Think of John Berryman's "Dream Song 366": "These Songs are not meant to be understood, you understand. / They are meant to terrify & comfort."[52]) What warm-blooded human doesn't suffer from feelings of inadequacy, fear, dashed dreams, a broken heart, panic, helplessness – from addiction, from pride, from hypocrisy and anxiety?

The trouble is that to make Plath's point to her mother, she not only doubles down on the idea that her pain can be compared to that of death camp survivors, but that she speaks for the survivors in defending her case. This is unacceptable in today's understandings of social justice. No one can speak for anyone else, certainly not about their suffering. And it's hard to imagine a more outrageous take than a middle-class white woman safe in England telling us what a Holocaust survivor went through, and what they want from art.

Perhaps the judgment of Rose, Hoffman and other Plath defenders is too lenient. Or perhaps here we find the wedge between Plath's modernist impulse toward universalism, on one hand, and on the other hand, postmodern assumptions that there is no bridging the gap across experiences (no grand narratives), no way of comparing, speaking for, each other's pain, to say nothing of speaking about *the* human condition.

While standing against fatalism of all kinds, both catastrophic and liberatory versions, I'm confident in predicting that today's open future will become the history of human emancipation only after the reclamation and cultivation of the impulse toward universalism. It will have to be a universalism rooted in the plight, knowledge and needs of the most disadvantaged, what Ato Sekyi-Otu, drawing on ethical frameworks indigenous to the African continent, calls "left universalism."[53] From this view, we see that the universal finds expression in particulars; and particulars, in relation to each other, with all their differences, contradictions and apparent separateness, constitute the universal. I'm reminded of Mohsin Hamid describing the two Sufi "paths to transcendence: one is to look out at the universe and see yourself, the other is to look within yourself and see the universe."[54] Part and whole are in constant (internally-related) processes of change, the outcome of which cannot be known in advance, mythologies of doom be damned.

NOTES

1   Sylvia Plath, February 4, 1963, in *Letters Home: Correspondence 1950–1963*, ed. Aurelia Schober Plath (HarperCollins, 1975), Kobo, italics added.

2   Plath, October 25, 1962, in *Letters Home.*

3   Heather Clark, *Red Comet: The Short Life and Blazing Art of Sylvia Plath* (Knopf, 2020), 871–75.

4   Sylvia Plath, "Edge," in *The Collected Poems* (HarperCollins, 1981), 224.

5   Plath, October 16, 1962, in *Letters Home.*

6   Janet Malcolm, *The Silent Woman: Sylvia Plath and Ted Hughes* (Vintage Books, 1994), part 1, chap. 8, Libby.

7   Anne Sexton, "Wanting to Die," in *The Complete Poems* (Houghton Mifflin, 1981), 142, italics in original. Sexton killed herself on October 4, 1974.

8   Melissa Bradshaw, "A Great Many Plathitudes: The Doom Myth of Sylvia Plath," *Quietus*, February 10, 2013, https://thequietus.com/culture/books/sylvia-plath-fifty-year-anniversary/.

9   Emily Van Duyne, "No One Gets Sylvia Plath," *LitHub*, November 6, 2020, https://lithub.com/no-one-gets-sylvia-plath/.

10 Clark, *Red Comet*, xix.

11 Marjorie Perloff, "The Two Ariels: The (Re)making of the Sylvia Plath Canon," *American Poetry Review* 13, no 6 (1984): 10–18.

12 Al Alvarez, "Poetry in Extremis," in *Sylvia Plath: The Critical Heritage*, ed. Linda W. Wagner (Routledge, 1988), 55.

13 Robert Lowell, foreword to *Ariel*, by Sylvia Plath, ed. Ted Hughes (Harper and Row, 1966), xiii.

14 Elizabeth Anne Roodhouse, "Re-Writing the Plath Myth: Sylvia Plath and the Cult of Celebrity in Print Publication" (Master's thesis, Department of English, University of Virginia, 2006), 12.

15 Anne Stevenson, *Bitter Fame: A Life of Sylvia Plath* (Houghton Mifflin, 1989), 33.

16 Van Duyne, "No One Gets Sylvia Plath," italics in original.

17 Robert C. Solomon, "On Fate and Fatalism," *Philosophy East and West* 53, no. 4 (2003): 435.

18 You'll find examples in any of Auster's many novels and in his delightful collection of haunting tales, *The Red Notebook: True Stories* (New Directions, 1992). Auster's own dense, lyrical, biographical articulation of his project appears in "The Book of Memory," the sec-

ond part of *The Invention of Solitude* (Penguin, 1982). In 2004, Auster provided a more accessible version during his interview with Eleanor Wachtel on CBC's *Writers & Company* (https://www.cbc.ca/radio /writersandcompany/paul-auster-on-writing-radical-life-changes-and -leaning-into-random-chance-1.7200314). In response to obituaries mischaracterizing Auster's interest in chance and coincidence as fanciful, reflecting a passion for melodrama, Siri Hustvedt (Auster's widow) asserted that Auster's core interest was always in "the complex patterns of often unpredictable interactions that affect every life" in a brilliant recollection called "The Mechanics of Reality," *LitHub*, May 28, 2024, https://lithub.com/remembering-paul-auster/.

19 John Jeremiah Sullivan, interview by Max Linsky, host, *Longform*, podcast, episode 585, June 26, 2024, 14:10, https://longform.org /posts/longform-podcast-585-john-jeremiah-sullivan.

20 Epictetus, *The Discourses* 1.12.23–24, Loeb Classical Library, https://www.loebclassics.com/view/epictetus-discourses/1925/pb _LCL131.91.xml?readMode=recto.

21 Niccolò Machiavelli, *The Prince*, in *Machiavelli: The Chief Works and Others*, vol. 1, trans. Allan Gilbert, 3 vols. (Duke University Press, 1956), 92.

22 Paul Dubois, *The Psychic Treatment of Nervous Disorders: (The Psychoneuroses and Their Moral Treatment)*, trans. and ed. Ely Jelliffe Smith and William Alanson White (Funk & Wagnalls, 1909), 240–41.

23 Paul Auster, *Oracle Night* (Picador, 2003), 45.

24 Perloff, "Two Ariels," 14.

25 Sylvia Plath, "Wintering," in *Ariel: The Restored Edition: A Facsimile of Plath's Manuscript, Reinstating Her Original Selection and Arrangement* (HarperCollins, 2004), 90.

26 Elisa Gabbert, "The Intolerable, I Guess," in *Any Person Is the Only Self: Essays* (Farrar, Straus and Giroux, 2024), Kobo.

27 Clark, *Red Comet*, 861.

28 Lillian Crawford, "Sylvia Plath: Will the Poet Always Be Defined by Her Death?," *BBC Culture*, July 20, 2021, https://www.bbc.com /culture/article/20210720-sylvia-plath-the-literary-icon-destined-to -remain-an-enigma.

29 Van Duyne, "No One Gets Sylvia Plath."

30 Gabbert, "Intolerable."

31 Derek Lam, "The Phenomenology and Metaphysics of the Open Future," *Philosophical Studies* 178, no. 12 (2021): 3898.

32 Lam, "Phenomenology and Metaphysics," 3896n2.

33 I feel obliged to note that I'm aware that my argument against fatalism sits in tension with my fear of uncertainty. As I explain in a later essay about parenting amid the eco-crisis, this fear of mine is so persistent and severe that I'd call it character-defining. Agonizing over both the problem of fatalism and the unknowability of the future might make me seem to be a glutton for anxiety. Is it fixity or fluidity that keeps me awake at night? Pick one! But the two aren't mutually exclusive problems. They are different expressions of what Elisa Gabbert calls humanity's "anxious relationship to time." See Gabbert, "Weird Time in *Frankenstein*," in *Any Person Is the Only Self.*

34 See Crawford, "Sylvia Plath." Note that Heather Clark urges us to read "Edge" not narrowly, as a message to Hughes, but expansively, as an indictment of "a sexist culture and literary tradition that equated perfect womanhood with passivity and compliancy," *Red Comet*, 875.

35 David Runciman, host, *Past, Present, Future*, podcast, season 4, episode 55, "The History of Freedom w/ Lea Ypi: Kant, Enlightenment and Peace," April 7, 2024, https://podcasts.apple.com/gb/podcast/the-history-of-freedom-w-lea-ypi-kant/id1682047968?i=1000651658070.

36 Runciman, "History of Freedom w/ Lea Ypi," 52:23.

37 John Naughton, "Aldous Huxley: The Prophet of Our Brave New Digital Dystopia," *Guardian*, November 22, 2013, https://www.theguardian.com/commentisfree/2013/nov/22/aldous-huxley-prophet-dystopia-cs-lewis.

38 Lea Ypi, in Runciman, "History of Freedom w/ Lea Ypi," 53:04.

39 Paul Auster, interview by Harriett Gilbert, *World Book Club*, podcast, "Paul Auster – New York Trilogy," *BBC*, November 3, 2012, 25:47, https://www.bbc.co.uk/programmes/p0104h52.

40 Rebecca Solnit, *Hope in the Dark: Untold Histories, Wild Possibilities*, 3rd ed. (Haymarket Books, 2016), xiv.

41 Plath, "Daddy," in *Ariel: The Restored Edition*, 75.

42 Plath, "Daddy," 76.

43 Jacqueline Rose, *The Haunting of Sylvia Plath* (Harvard University Press, 1992), 228.

44 Plath, "Lady Lazarus," in *Ariel: The Restored Edition*, 16–17.

45 Rose, *Haunting of Sylvia Plath*, 206.

46 Plath, October 21, 1962, in *Letters Home*.

47 Theodor Adorno, "Cultural Criticism and Society," in *Prisms*, trans. Samuel M. Weber (Neville Spearman, 1967), 34.

48 Rose, *Haunting of Sylvia Plath*, 207.

49 Alan Williamson, *Pity the Monsters: The Political Vision of Robert Lowell* (Yale University Press, 1974), 4.

50   Steven K. Hoffman, "Impersonal Personalism: The Making of a Confessional Poetic," *ELH* 45, no. 4 (1978): 699.

51   Hoffman, "Impersonal Personalism," 700.

52   John Berryman, "Dream Song 366," in *The Dream Songs* (1969; Farrar, Straus and Giroux, 2014), 388.

53   Ato Sekyi-Otu, *Left Universalism, Africacentric Essays* (Routledge, 2018).

54   Mohsin Hamid, "Are the New 'Golden Age' TV Shows the New Novels?," in *Discontent and Its Civilizations: Dispatches from Lahore, New York, and London* (Riverhead Books, 2015), 120.

CRISIS MOVES

"COVID-19 Pandemic Hastens Canada's Urban Exodus."
– Julie Gordon, *National Post*

"Survey Finds Moving Is More Stressful for Many than Getting Divorced, Having Kids."
– *CBS News*

People's lives, in Jubilee as elsewhere, were dull, simple, amazing and unfathomable – deep caves paved with kitchen linoleum.
– Alice Munro, *Lives of Girls and Women*

About two weeks before Jess and I bought a house in the small town of Paris, Ontario, I stood on the lawn of our home in Hamilton and said we'd never leave the city. I was talking to a contractor we'd asked to estimate the cost of converting our uninsulated garage into year-round livable space. Turned out the contractor grew up in Lanark County, not far from the small town where I was raised. After sharing news about mutual acquaintances, he asked how Jess and I liked living in Hamilton. I said, "We wouldn't want to be anywhere else."

I meant what I said when I said it. We lived in a yellow-brick, one-and-a-half-story detached home in the city's fast-gentrifying east end. I loved our neighbours, the large maple tree in our front yard and the squirrels popping in and out of the softball-sized hole in the tree's trunk. Jess and I moved into the house in August 2017, when she was pregnant with Gus. In February 2021, Winnie was born in the basement. Both kids' placentas are buried under trees in the garden, symbols of our rootedness in a place we thought we'd live in forever. But on January 15, 2022, we moved fifty kilometres west, from a city with a population of 771,000 to a town of around 15,000 people.

By relocating when we did, my family joined the ranks of pandemic movers. I'd heard stories about people fleeing cities from the early days of the Covid-19 lockdown. Rich families steering clear of crowds by moving to their summer homes. College students leaving cramped dorms for their parents' places. In December 2020, a *Fashion* magazine profile of pandemic movers began: "If it feels like every other friend is ditching downtown for the suburbs or somewhere greener, it's not just you. This year saw droves of young urbanites chasing manageable rents (and mortgages), more physical space and a slower pace of living."[1] An article in *Time* comparing pandemic-driven moves to moves in the midst of the sixteenth-century plague noted: "2020 is by no means the first time wealth has enabled mobility in a time of crisis."[2]

Between 2020 and 2021, more people moved out of Toronto than into it. In the US, permanent moves increased 15 percent in March 2020 over the same month a year earlier. In cities such as Houston, Texas, moves spiked more than 60 percent year over year. In the Hudson Valley, north of New York City, traffic jams formed around houses for sale on rural roads. In 2022, the *Globe and Mail* announced: "Canada's urban exodus picked up steam into the second year of the COVID-19 pandemic."[3] Elizabeth Strout's *Lucy by the Sea*, shortlisted for the Booker Prize in 2022, tracks one family's pandemic move from New York City to a small town in Maine.

Decades from now, when young people ask what it was like to live through the Covid-19 pandemic, those of us who were around will talk about feeling as though the world suddenly stopped. School closures, borders sealed, factories shuttered, travel restrictions. According to an Organisation for Economic Co-operation and Development (OECD) report, "By spring 2020, more than half of the world's population had experienced a lockdown with strong containment measures."[4] Yet in the middle of this period of historic immobility, some of us picked up and moved. In the case of my family, one crisis (of global health) begat another (of feeling out-of-place; that is, failing to feel what social geographers call "at-homeness"[5]). Under the stress of these compounded crises, I developed a habit of what I describe in this essay as a form of "crisis thinking." Crisis thinking distorted my sense of place, time and identity for several months on both sides of the move. At the time, these routine flights of fancy felt singular to me; but upon reflection, I wonder whether most people moving from one place to another are prone to similar distortions of mind (to greater and lesser degrees). If it's true that crisis thinking is commonly triggered by moving, that would help understand why moves are often so distressing, even among those who move by choice, even in the best of circumstances.

The writer Anne Thériault told me that before the pandemic she'd never considered leaving Toronto. She'd lived in a midtown apartment for fifteen years. Her son grew up there. She'd built strong friend networks in the city and loved living where "people are always coming from other places."[6] Anne thought of herself as "a big city person."

The Covid-19 lockdown quickly changed her outlook. Anne said to herself, "I can't imagine having two adults working from home in this apartment, plus a kid doing school from home and no outdoor space." She and her son began staying with her mother in Kingston, to have access to a yard and more indoor space. Returning to Toronto was "very stressful," especially before vaccines were available. Anne laughed sheepishly while describing her Covid nightmare: "There's this scene in *Little House on the Prairie* where they all get malaria, and everyone's really sick at the same time, but they can't take care of each other. My nightmare was: We're all going to get Covid, all at the same time, and I'm not going to be able to take care of my kid because we're all going to be too sick."

In spring 2021, Anne and her husband knew that they would "go crazy if we stayed in the city." They began looking at houses for sale in Kingston. They offered to buy a place downtown but, considering the fiery-hot housing market, didn't expect anything to come of the bid. Twelve hours later their realtor called to say that their offer was accepted. "We said, 'Oh shit! I guess we're moving to Kingston!'"

Our move to Paris did not follow the logic of Anne's story. Like everyone living under lockdown, we were sick of our four walls. We were highly Covid-cautious – no stores or playdates – which, with two kids under the age of four, meant living with a special kind of chronic cabin fever. But prior to stepping into what would become our new home, not once did we talk seriously about leaving Hamilton.

In September 2021, the city was gradually reopening, and Jess and I were twice-vaccinated. We were talking with several building companies about adding a bedroom to the house. But, we told ourselves, before sinking major resources into home renovations, responsible people would consider every option. We couldn't afford to move within our Hamilton neighbourhood. What about looking at something closer to my campus, which would also put us nearer Jess's family? A nice big house in a cute small town. Didn't we joke about envying the people on *Gilmore Girls*? Of course, we wouldn't actually move. Just doing due diligence.

Was ours even a "pandemic move," then, in anything other than the sense that we moved within a few years of the outbreak of Covid-19? We didn't experience pandemic pressures building to a breaking point; didn't form a decision to move, then hunt for houses. Didn't, as Anne's family had, spend time in potential new hometowns. But while "what-ifs" are not evidence, it's difficult to imagine that we would've ended up in Paris if we'd not spent the last two years fearing crowds, pining for new experiences, virtually cut off from the things we loved about our city. Under the pandemic, we'd cultivated self-sufficiency, watched friends on our street leave Hamilton for more green space and talked periodically about what we'd do differently during the plague next time around. Rebecca Mead's *Home/Land: A Memoir of Departure and Return* describes her family's move from Brooklyn to London in the aftermath of Trump's 2016 election win. To borrow her phrase, our "voluntary, privileged exile"[7] from Hamilton was conditioned by the pandemic.

An hour after the realtor told us that the house in Paris was ours, Jess lay on our bed in the dark, sobbing while I held her head. What if we felt isolated, got bored, made no friends, never adapted to our new surroundings? What if small-town kids rejected Gus's budding queerness? Have we made a terrible mistake?

You'd think we'd have answered these questions before buying the house. Well, we had. In the twenty-four hours between seeing the house in Paris for the first time on Thursday afternoon and submitting our offer to buy on Friday afternoon, we did nothing but debate the pros and cons of moving. The main pro was the Paris house itself. Bigger (two living rooms, a larger foyer, a bedroom for both kids plus an office for me), on a corner lot, with a park across the street and an elementary school beyond the park. High-ceilings, a newly-renovated kitchen, dark acacia floors throughout the main level and the original (circa 1910) pine flooring upstairs.

Living in Paris would mean we were "closer to nature," and that would be a good thing. I talked about "getting into hiking," and pictured our family in a canoe, paddling down the Grand River. On the podcast *Townsizing* Noelle Murrain says she's considering moving to a small town for "just the slower pace. [. . .] There's something about the energy of being in a city, where, the minute you go outside you already feel tired. Too much stimulation."[8] As a recovering alcoholic, I've often been agitated by drinking culture in the city. The thrill and stress of crowds in the afternoon reminding me of the pleasure of a drink (or fifteen) after work. Summer sidewalks lined with patios smelling of cold beer. Booze ads blanketing the visual landscape. I told myself the slower pace in a place *Harrowsmith* magazine once called the "Prettiest Little Town in Canada" would bolster my sobriety.[9] Besides, moving would be an adventure! The pandemic deprived us of so many experiences we wanted to share with each other and the kids. The monotony of months under Covid-19 lockdown left us hungry for something new. Like Mead's family, "We were ready to be set in motion, to veer from what seemed to be the predictable path of our lives."[10]

There were only two cons. First, we loved our home in Hamilton. We lived two blocks from a great public school. We were a twenty-minute drive from downtown, and the beach, and the train station to Toronto, and the botanical gardens with their dozens of kilometres of trails.

The other con was what we imagined would be the limits, if not problems, of living in a small town. Fewer restaurants, shops and unexpected street encounters, sure, but we refused to consider that sort of thing a significant loss. We should cook at home more often and buy less crap anyway. And what exactly was the *uniquely urban unexpected encounter* that we would regret missing? Action on yet another condo construction site? The guy at 11:00 a.m. yelling "GET BACK HERE YOU COCKSUCKER!" at his enormous dog charging down the sidewalk on Ottawa Street? There would be unexpected encounters in Paris, just different ones.

What about the adage: small towns, small minds? Hamilton is a diverse city of nearly a million people. Paris is a largely white town of a few thousand. On the same episode of *Townsizing* with Noelle Murrain, a different guest considering a move worries about "the homogeneity of a small town."[11] Murrain herself, a Black woman, says when shopping in a town "and I don't even see hair products that are for me? Then that tells me something about the community that I'm in." (There are no products for Black hair in the drugstore in downtown Paris.) The County of Brant routinely elects Conservatives to legislative office. What sort of boy or girl did we imagine Winnie dating in high school? Did they even have lesbians in Paris? With being "closer to nature" comes fishing, hunting and ATVs, the stuff of redneck masculinity.

In her memoir *Open House: A Life in Thirty-Two Moves*, Jane Christmas realizes that Brixham, a small town on the British coast, is not what she dreamed it would be when moving there: "I thought I would be happy living a small-town life, but I was not. I felt cut off from the wider world. I wanted more: more culture, more discourse, more diversity, nicer shops, galleries, a cinema, more anonymity, and I began to not only resent the absence of these things but mourn them like the loss of oxygen. Each time we drove back into town after being away for a few hours or a few days I could feel a part of me dying inside."[12]

After the sellers turned down our first offer, we visited Paris a second time to decide whether to increase our bid. It was a perfect early-fall Saturday: blue sky and sunshine, but cool enough to wear long sleeves. Entering the heart of Paris from Rest Acres Road takes you down a long, curved slope into the Grand River Valley. From a few hundred feet above, you see the red-bricked storefronts of Grand River Street North, edged by the river to the east. A curtain of white water falls over Penman's Dam. A tall iron bridge spans the river, carrying lumbering cargo trains every few hours. You see the yellow bricks of the Arlington Hotel; the spire on the church built in 1887; the dome on the central library; the restored textile mill, now a bustling market; and homes scattered among trees and small streets, including the home we were there to consider buying.

Who wouldn't want to live in this storybook village?

That day we spent in Paris lived up to the promise of our opening, idyllic view from the car. The house was even brighter in full daylight. The ceilings were as high as I'd remembered. Gus marched around the front yard with a pencil and notebook, seemingly already taking possession of the property. We met the neighbours, lovely people (one a successful artist), who moved to Paris thirty-three years ago, when their kids were small. We searched for flaws, but the ones we discovered were easily fixable – repainting a room here, filling gaps in a handsome baseboard there.

Downtown, we walked through two bookstores, a tea shop and an independent shoe store. Even the gross businesses – realtors, banks and insurance companies – were housed tastefully in stone buildings. Eating lunch on a patio overlooking the Grand River, I was pleased by what I saw in other diners. A young dad wearing retro New Balance shoes. A mixed-race family in boating gear. Lots of smiles, lots of kids, lots of outdoor activity-wear. I'd happily be neighbours with these people. Before we left the restaurant, I phoned our realtor and directed her to issue a second offer.

Through the winter and spring after we moved to Paris, I often thought back to that October lunch on the Grand. By July, it was clear to me that the people I'd imagined as my new neighbours last fall must've, in fact, been tourists. I'd never seen any of them in town again and living in Paris through the summer taught me that virtually every person not doing paid work downtown is visiting from somewhere else. Many tourists live only a short drive away: Cambridge (twenty minutes) or Hamilton (forty minutes). Many more have driven a few hours, from Toronto and its suburbs. Large tour buses frequently stop in Paris for an afternoon.

When recalling my buoyant feelings of that October lunch, I feel silly. Not because I didn't end up getting to be friends with cool-shoe-dad and mixed-race-boating-family, but because of how badly I'd misunderstood the character of the town I was visiting for the express purpose of learning about it. My conception of moving to Paris looked a lot like this day: sunny and clear, small streets bustling with diverse crowds, eating and playing by the Grand and the Nith (Paris's second, but no less charming, river).

That night, our realtor called after the kids were asleep to tell us our second offer was successful. I felt obligated, for her sake, to express unbridled joy, but when I looked for such emotion in Jess's face, I saw instead that she was holding back tears. Shocked? Too happy to speak? No, this was the moment the fun fantasy of living in Paris was overtaken by the painful reality that we were moving. Perhaps the crisis of our move erupted in this moment – in the sense that the deal launched us into a period of emotional free fall. Or perhaps the decisive turning point – move? don't move? – having now been passed means the crisis was resolved, and our emotional turmoil was post-crisis trauma.

In any case, almost immediately we lost hold of all the best-case projections of our new life, and accepted all the worst possibilities as inevitable. We'd ensured our kids wouldn't go to school in classrooms full of kids from other progressive families. We'd killed off the handful of friendships Jess had built with neighbourhood moms over the previous three years. What about the kids' placenta trees? We'd thrown ourselves from stability and happiness into disorder and regret. Melody Warnick, author of *This Is Where You Belong: Finding Home Wherever You Are*, would say we demolished our cherished sense of "place attachment" without preparing ourselves to live among the ruins.[13]

We began staying up till 2:00 a.m. each night getting the Hamilton house ready for sale. I'd wake up two or three hours later and begin again: emptying cupboards, painting door frames, scrubbing the furnace room floor. Once, around midnight, I nearly fell off the garage roof while patching a seam by the skylight. We spread woodchips in the backyard where the kids' pool killed the grass. We filled a storage unit with furniture, pressure-washed patio stones, hired a stager, hung decorative lights, had the entire inside of the house painted.

I lost ten pounds in two weeks. I stopped doing all but the most essential parts of my job, which meant no longer working on this book, which had been going well through the first three-quarters of the year. In fact, at the time our sudden decision to move up-ended our lives, I was working on an essay about my midlife crisis of a few years earlier, an essay which carries the suggestion that such drama is behind me. My anxiety and paranoia were such that I wondered whether my subconscious drove me to move in order to provide material for the book.

By the time we sold the Hamilton house in mid-October, I'd broken out in a red, bumpy rash around my waistline so painful and itchy, I assumed it was a reappearance of shingles. Except, shingles typically manifests on only one side of the body. This rash wrapped around me like a belt. After examining my skin and hearing my story, my doctor called it a "stress rash."

In her book *On Moving*, Louise DeSalvo says that moving from a house she loved for thirty years produced "a sense of loss almost as profound as when my mother died a few years before. Surely a move shouldn't feel like mourning the passing of a family member. But it did."[14] My rash cleared up before our closing date, but the tumult of the move – the sadness, fear and confusion it created – lasted throughout our first year in Paris.

The pain, a mix of grief, fear and disbelief, did something very strange to my perception of time. For a while, the abrupt decision to make such a major change made me exaggerate the significance of every minute, every experience. It was obvious that moving changed the course of our lives fundamentally. The question was whether the change would be for the better or worse. I began as-sessing the value of the change based on the details of whatever particular moment I was in. It might sound like a classic case of what psychologists call catastrophist thinking – that is, "to view or

---

present situations as considerably worse than they actually are."[15] Catastrophizing means jumping to "the worst possible conclusion, usually with very limited information or objective reason to despair."[16] I certainly did plenty of that. But I could just as easily (and just as often did) use very limited information to view situations as considerably *better* than they actually were. I fabricated visions of future fulfillment or desperation – only extremes were imaginable – based on each passing encounter.

For example, on our first family walk past the old factories in Paris, we met a guy carrying a tailless orange cat. It wasn't a Manx, he said, just a rough furry friend who'd lost his tail in an accident. When I was a kid, I told the guy, I had a cat who'd lost a tail after being hit by a hay baler. Thor was his name. "Thor?" the guy said. "That's funny. Mine's named Loki. That's Thor's brother in the Marvel universe." I took it as a sign that we'd landed in the right place.

But two months after we moved, a black pickup truck caked in mud parked on the street in front of our new house. A neighbour? White words on a black flag flying above the driver's door read: "Don't F&#*@n Touch It!" I imagined Gus fifteen years from now with a truck like this one, or five years from now coveting a truck like this one. Oh god, I've put my kids in harm's way.

But then an old friend visiting our new home remarked on how lucky we were to escape Hamilton's toxic air. He's convinced that living close to the steel plants caused his daughter's asthma. *We're escaping*, I thought to myself. Decades from now, my kids' lungs will be pink and hale because of the move I'm making today.

One Saturday in late November, six weeks before our move, we drove to Paris midafternoon to check out a Christmas market. Some kind of no-tax downtown local thing promoted by the business association. There would be a tree-lighting ceremony. Stores would be open late.

The trip was a total bust. Shops weren't open in the evening like we thought they'd be. No bustle. No foot traffic. We're walking from dim storefront to dim storefront. Quiet. How soon will I get bored of the corner of Grand River and William Street? I'm already bored of it, aren't I? The shops along that strip are for tourists. How many chocolatiers, high-end furniture and jewelry shops will I need in a typical month? We ambled along, Gus refusing to walk, Winnie wailing, Jess cold. Nowhere to go, feeling stuck. I thought to myself: We will be stuck here forever.

But next time I was in town, a stranger smiled at me at the Dog-Eared Café, and I sensed that moving to Paris would be fine. Cozy, friendly, fulfilling.

My thinking made every moment a crisis: the decisive split between this move being a good or a bad idea. My hopes for the future rose and fell with each unique experience. I was trying to foresee (predict, anticipate, imagine, envision) a life based on extremely limited information.

My inclination to do so is understandable. Experience is evidence we use to envision our future and make decisions we hope will lead to good things. Yet even at the time I knew my particular habit of mind was flawed. Because trustworthy predictions are informed by trends. And limited information makes identifying trends very difficult, if not impossible.

Scientifically speaking, it's deeply problematic to make predictions based on a "small n" dataset. The variable "n" denotes the number of cases being analyzed. Let's say there are 15,000 people in Paris. To make a statistically significant assumption about the population (with a margin of error of 2 percent, correct 95 percent of the time), you'd need something like 2,070 cases. The "n" would have to be 2,070. That's not the way our lives work, I get that. We don't calculate the probability of friendships forming, taking a pleasant walk, being dazzled by a sunset or enjoying takeout food based on statistically significant samples. Yet, for more than six months after our move, I made solid predictions about what our lives would be like based on an "n" of one.

The streets bustled with people the day we looked at the house. *It's a cute little town, with a busy street life.* There was a beautiful sunset. *There will be beautiful sunsets every night, and we will enjoy the beautiful sunsets.* Two hip-looking families sat next to us at lunch. *There are cool people here, our age. We'll be part of a growing community of an artistically-minded, nature-loving community of cool people.* The Christmas market was a bust. *Oh my god. This place is a ghost town. There is no street life here.* At Lions Park, the first three families we meet are from out of town. *This place is just a tourist playground. There's no homegrown community.*

This tendency to envision the future as the long-term extension of the present moment is a form of crisis thinking. Projecting a life story from a small "n" transforms a moment into the decisive turning point that leads one way or another. None of the particular experiences I placed such stock in could even remotely be considered actual, objective crises. (We had our health and our stuff. We weren't fleeing violence. We'd actually made money on the real estate transactions.) But the subjective component – that is, my approach to each experience – overwhelmed reality, turning each moment into either fuel for catastrophism or for starry-eyed relief and promise. But either way, interpreting each experience as a microcosm of the story of the rest of our lives imbued it with far more meaning, higher stakes, than was reasonable. My experience of the move was defined by crisis thinking from the moment we first saw the Paris house (when I decided this was our once-in-a-lifetime chance to take up a place in the small-town idyll), until nearly a year after we'd moved (through much anguish and remorse).

The move itself was frigid. Men in boots tracking snow and salt through two houses. Half of our plants died in cold moving trucks. My big orange tabby shat in his cat carrier riding next to me in the car from Hamilton to Paris. An omen? Those first weeks in Paris, I saw omens everywhere. Worst was what I found in the attic. Kneeling and feeling for drafts by a small window, I saw bones lying on the floor next to me. They comprised a full skeleton. It was as though the skeleton had been picked clean and preserved for a science class. Not a bone missing or chipped. No rotting flesh or feathers attached. A bird? Squirrel? Baby raccoon? An offering to dark gods left by previous owners? I couldn't tell.

In Paul Auster's *Winter Journal* he talks about finding a dead crow in the house he and Lydia Davis lived in in upstate New York during their short, doomed marriage. Auster doesn't believe in ghosts, but feels that their house is haunted. In the crow, he saw "the classic omen of bad tidings" and their marriage collapsed within a year of finding it.[17] I didn't have a bag or anything to put the bones in, and I didn't want Jess or the kids to see them, so I left them lying where I'd found them for a few days. The skeleton lay on the floor of the attic, ten feet above our bed. Before falling asleep, I lay there thinking of it lying above me.

Our first months in Paris were isolating. It was the coldest, snowiest winter in memory. We couldn't really go outside even if we'd wanted to because it was the peak of the Omicron variant of Covid-19. Nearly everything about the move we'd been excited about – hikes, building community, exploring the town – felt unattainable. While the thing we dreaded most – feeling physically and socially isolated – was now our daily reality.

To make matters worse, there was all kinds of trouble with the house itself. Our first night in the home, the pipes connected to the bathtub leaked through the ceiling onto the basement stairs (talk about omens). The upstairs toilet moaned for fifteen seconds after every flush. We called it the "Paris-saurolophous" – because Gus loves dinosaurs, and the Parasaurolophus is the dino with the crest on its head that honks. You flush the toilet and wait. Then yell, "A Paris-saurolophus is running down the street!" while smiling at Gus and wondering what will break next.

The washing machine ran only hot water. Gus's room was very cold, despite our covering the window in plastic. The door to the basement wouldn't close. Neither would the bedroom door. Gus's door didn't have a handle. The gas fireplace stank. The whole main floor stank (of gas?). The bathroom sink faucet sprayed water on your clothes when running at normal pressure. Hammering a nail into the upstairs hall, I heard something large drop inside. Was this a normal level of move-in trouble, or had we bought a lemon? I imagined being interviewed for a TV show about people who unwittingly move into wrecks. Then the basement flooded.

Standing in the kitchen in morning darkness, bleary-eyed, waiting for coffee to brew, I heard unidentifiable splashing. With Gus in my arms, I ran down the basement stairs to see water gushing through the stone foundation. It had rained overnight for the first time since we moved. Snow melting in the yard poured into our utility room. In one spot, water was coming through the wall so rapidly, with such force, that it spurted into the room as if from a backyard garden feature. Weeping walls, a crumbling foundation – the contrived symbolism, the on-the-nose pathetic fallacy of an unimaginative poet. Was our house about to collapse? We were falling apart.

Rebecca Mead writes that her family "chose movement" in the opening years of Trump's presidency "because movement is a kind of freedom, too."[18] What about when freely chosen moves kick off crisis? How are these crises of moving resolved?

Resolution is a troublesome concept. If we follow a standard dictionary definition of "crisis" as a temporary period of extreme trouble, crisis is characterized by unsustainable instability, and, therefore, imminent transformation. We'd say that the crisis is resolved when the situation stabilizes, when sustainable ground, a new normal, emerges. When precisely does that occur? Talking about crises being resolved can imply too pat a conclusion to situations that are, by definition, messy, tempestuous, unpredictable. And yet, crises do end, temporariness being another of their defining features.

Some pandemic movers experienced a sense of resolution immediately upon relocating. For example, in the case of a woman named Hope profiled in *Fashion*, moving from Toronto to Dryden, Ontario, in summer 2020 resolved underlying self-image problems. "I used to be driven by things like status and security and being in Toronto really brought out these insecurities in me. Being in Dryden, I've been able to take a step back and really commit to personal growth without the pressures of the race I was subjecting myself to. [. . .] Since leaving the city, I feel more together than I ever have."[19]

Marian Amo was born in Ghana and moved as a child with her family to Brooklyn, New York. In March 2020, Marian fled her small New Jersey apartment to live with her sister in West Virginia. In July, she decided to remain in the Mountain State permanently. She loved the topography, the outdoor activities available to her after work every day. In West Virginia, she found it easier "to reconnect with yourself and with nature. When I was in New York, after work, all you could do was just get home, watch TV, and go to sleep and do it again. In West Virginia, there are places to run to, explore, hiking trails. That was nice for me."[20] She misses the food in New York City. "But I don't miss the commute. I don't miss the crowdedness. I don't miss the subway smelling really bad."

Other pandemic movers not only failed to resolve the problems motivating their move but ended up falling deeper into crisis. For example, when Gabrielle Drolet's Toronto-based graduate program went online in 2020, she moved to the small town of Wolfville, Nova Scotia, to live with her girlfriend and enjoy more affordable housing. At first, the rural idyll delighted her: "Wolfville is cute storefronts and little cafes, trips to the Sunday farmers' market, and walks along the dykes – the mingled smell of salt and mud in the morning."[21] Over the year, though, especially during winter months, Drolet's mood fell, her "mental health faltered." Unable to drive much, she became socially isolated. Accessing necessary health services was difficult, often impossible. She envied friends in cities still able to gather in parks and enjoy small pleasures like bubble tea and Indian takeout. Of course, Drolet writes, moving to a small town can help solve financial problems and relieve stress for lots of people. It just isn't a healthy option for everyone; certainly not for her.

While Drolet doesn't blame people for moving to small towns to save money, she rightly points out that doing so resolves problems exclusively at the individual level. Socially, the urban exodus actually extends "the affordability crisis – it just moves it around, driving up the cost of living in suburbs and rural areas and making life more difficult for locals who've always been there."

I asked a neighbour and prominent community activist who's lived in Paris for ten years what changes she's seen here since the start of the pandemic. The rising cost of housing is her primary concern. The typical house in Paris's sprawling new subdivisions lists for over a million dollars. Relatively few immigrants, first-time homebuyers, minimum-wage workers and retirees can afford to buy such a home. "There's no housing for people to downsize into" in Paris, my neighbour told me.[22] She cheers the growing racial and ethnic diversity in town, another recent change. However, she fears that as more people with money arrive, the town will become more homogenous in some respects, less diverse socio-economically and generationally. The housing crisis in Toronto creates housing crises of different kinds in Paris, Brantford and other smaller southwestern-Ontario communities.

Although we didn't know it at the time, the flood in our new basement marked the beginning of the end of our moving crisis. We hired a team to waterproof the home's foundation and finish part of the basement. Between March and September of our first year in Paris, more than thirty different tradespeople worked on the house. They transformed a damp, grimy underground cave into cozy living space. From builders, we picked up home-maintenance tips, as well as stories about Paris, Brantford, Brant County and neighbouring areas farther afield. We became more trusting in the integrity of the house and more rooted in the town. Although we didn't try to replicate Melody Warnick's rigorous approach to putting down roots, we found ourselves conducting many of her "Love Where You Live experiments" – walking often, getting to know neighbours, exploring nature, volunteering (at local schools and arts organizations) and staying loyal (or at least not giving up) through hard times. In Warnick's words, we'd begun forging "the solidly affectionate bond of place attachment, with the rewards that went along with it: stable relationships, secure children, good health, and, for that matter, a nicer town."[23]

By the time our basement was renovated, we were regularly having pizza nights in the park with new friends. We'd learned the names of dogs whose daily walks passed our porch. Our kids worshipped our neighbours' grandkids, who often played outside. I was struck by the harmony between resolutions to the physical and mental problems of our home. We began feeling pride in our accomplishments – in the house, in our budding relationships. When the snows of our second winter melted, we felt relieved and strong (and dry).

Six months after the flood, we hosted a picnic in the park across the street from our house. It was an impromptu, informal gathering, thrown together after learning that I'd neglected to cancel our wedding photographer while cancelling our wedding months earlier (after having postponed it twice during the pandemic). Making lemonade out of very expensive lemons, we invited everyone we knew to have free professional photographs done while catching up with us and meeting new friends.

Over the afternoon, more than sixty friends and family members dropped by. Old neighbours from Hamilton laughed with new pals from Paris. Jess's family from Port Dover met guys I'd met in university twenty-five years ago. Gus and his best friend from our new street hoisted their toy monster trucks and yelled Cheese! in front of the camera. Winnie gently tapped a pinata with an old metre stick while older cousins stood by twisting in anticipation of their turn. Jess's grandma scolded kids from four different southern Ontario municipalities while they grabbed candy from a salad bowl at her feet.

At the photographer's command, Jess and I snuggled on our porch stairs, on our new couch, in the doorway to our kitchen. He gamely followed us down the banks of the Nith to photograph us standing in the shallow river whose meandering beauty was key to making Paris our home. In our basement, I pointed to the spot on the wall that gushed water in the rains of last spring. The hole was now patched and hidden by foam insulation and painted drywall. "You'd never know," said a friend.

On October 2, 2022, the anniversary of our buying the new house, it occurred to me that I'd now seen Paris in all seasons. Paris's blizzards, mud in March, the green canopy of Barker's Bush in June, the town's brown lawns in July – I'd seen them before. I had ridden the Ferris wheel at the Paris Fair on a hot Labour Day weekend. I'd felt the blend of emptiness and relief downtown the day after summer tourist crowds disappeared.

I told Jess that while I felt increasingly at home here, we would figure out how to return to Hamilton if she needed that to be happy. She surprised me by saying without hesitation: "I don't ever want to move again. I would miss this place." Perhaps our original vision of what the move could be as we viewed Paris from the top of the valley wasn't so far off the mark.

I take in that view nearly every day now. It's our drive home from Gus's school. I sometimes think of it as the view that sealed our fate. The train bridge over Penman's Dam. Traffic moving slowly on the main street below. Church spires, brick houses, the compactness, the coziness, the ideal of a small town. The view is more familiar now, a year later, but hardly less magical, no less inviting. I still look over the town every time I drive that way, trying to take everything in before moving out of view. But now I look for familiar landmarks. Where is our home in relation to that hill? Which street cuts that angle near the school? There's the path we take on our way to throw stones into the Grand. And I wonder what the view will look like years from now. Already you can see the bright skeletal lumber framing new houses in developments near the high school. How many new developments will you be able to see when Gus and Winnie are teenagers? How many trees will be lost?

But while I wish for the survival of Carolinian forests along the Grand, and I worry about subdivisions eating up farmland, this isn't crisis thinking. From the top of the town, I look over Paris happy, knowing that the future is uncertain, open-ended. Neither the joy nor trouble I feel here today holds the answer of what I'll feel years from now. Becoming rooted in Paris enriches the present, lowers the stakes of each moment. The crisis of the move, or the crisis-thinking the move set in motion, is behind me.

We tend to think of the pandemic years in terms of restricted movement. Lockdown, shelter in place, work from home. But the crisis also caused movement on a mass scale – people fled crowds, were forced from care homes, made snap decisions to relocate. I'm tempted to say that every move is a crisis, at least in miniature. Even the easiest, most desired move involves central aspects of the concept – a critical point, a crossroads, a decisive transformation. However, moves need not involve the intense danger associated with crisis, the potential for catastrophe, existential fear, though our move to Paris certainly did. And it's safe to assume that we're not the only people who experienced major emotional turmoil among the hordes who moved during the pandemic.

In October 2022, *Toronto Life* profiled five families who left the city under lockdown only to return within a year or so. "They missed their friends, their short commutes and their favourite haunts. They missed the city's energy, spontaneity and abundance. In short, they missed everything."[24] Five months later, *Fortune* magazine reported: "Three in four Americans say they regret something about their move in the past year."[25] Would the same be true if 2022 hadn't seen record-setting inflation and the first significant interest rate hikes in decades? When we consider the nested nature of crises — a personal move within a cost-of-living crisis within a pandemic within a global economic crisis — it becomes very difficult to distinguish causes from effects. Keith Gessen says it's too soon to know how growing up under the pandemic has affected kids: "It's going to take a lot of time to sort through all the damage."[26] I'd say the same about the social consequences of pandemic moves. I say it while watching robins repair a nest outside my office window. When they built the nest last spring, my view was unfamiliar and disorienting, like I was looking through someone else's eyes. Today, it looks like home.

NOTES

1   Jillian Vieira, "Why These Long-Time City Dwellers Moved Out During COVID-19," *Fashion*, December 11, 2020, https://fashionmagazine .com/flare/moving-out-of-the-city-because-of-covid/.

2   Suyin Haynes, "COVID-19 Is Prompting Wealthy People to Move Out of Cities. The Plague Had the Same Effect Hundreds of Years Ago," *Time*, August 21, 2020, https://time.com/5851978/pandemic -plague-henry-viii/.

3   Thomson Reuters, "More People Leaving Toronto, Montreal for Smaller Pastures as Pandemic Hastens Urban Exodus," *CBC*, Janu-

ary 13, 2022, https://www.cbc.ca/news/canada/urban-exodus-canada
-toronto-montreal-Covid-19-1.6313911.

4   Dorothée Allain-Dupré et al., *The Territorial Impact of COVID-19: Managing the Crisis Across Levels of Government* (Organisation for Economic Co-operation and Development, November 10, 2020).

5   Elena Ariel Windsong, "There Is No Place Like Home: Complexities in Exploring Home and Place Attachment," *Social Science Journal* 47, no. 1 (2010): 205–14.

6   Anne Thériault, interview with author, Zoom, March 7, 2023.

7   Rebecca Mead, *Home/Land: A Memoir of Departure and Return* (Knopf, 2022), chap. 1, Kobo.

8   Anne Helen Peterson, host, *Townsizing*, podcast, episode 6, "Moving to a Small Town: The Pros and Cons," HGTV, November 21, 2022, 28:13, https://townsizing.simplecast.com/episodes/moving-to-a-small-town-the-pros-and-cons-kaR2qnOL.

9   Quoted in Robynne Trueman, "Ontario's 10 Most Charming Small Towns to Visit This Summer," *Travel*, August 6, 2022, https://www.thetravel.com/small-towns-in-ontario/#stratford.

10 Mead, *Home/Land*, chap. 1.

11 Laura Wennstrom, quoted in Peterson, "Moving to a Small Town."

12 Jane Christmas, "The Acknowledged Catalyst," chap. 1 in *Open House: A life in Thirty-Two Moves* (HarperCollins, 2022), Libby.

13 Melody Warnick, *This Is Where You Belong: Finding Home Wherever You Are* (Penguin, 2016), 16.

14 Louise DeSalvo, introduction to *On Moving: A Writer's Meditation on New Houses, Old Haunts, and Finding Home Again* (Bloomsbury, 2009), Kobo.

15 Ricky Green, "Catastrophizing Life's Problems: On the Relationship Between Attachment Anxiety and Belief in Conspiracy Theories," (PhD diss., University of Kent, 2021), 19.

16 "Catastrophizing," *Psychology Today*, no date, https://www.psychologytoday.com/ca/basics/catastrophizing.

17 Paul Auster, *Winter Journal* (Faber and Faber, 2012), 90.

18 Mead, *Home/Land*, chap. 1.

19 Vieira, "Why These Long-Time City Dwellers."

20 Marian Amo, interview by Christi Cassidy, host, *Moving Along*, podcast, episode 8, "From Brooklyn to West Virginia – Pandemic Moves," March 10, 2022, 35:56, https://movingalongpodcast.com/from-brooklyn-to-west-virginia-pandemic-moves/.

21   Gabrielle Drolet, "The Pros and Cons of Moving to Small-Town Nova Scotia During the Pandemic," *CBC News*, March 11, 2022, https://www.cbc.ca/news/canada/nova-scotia/unlocked-creator-network-wolfville-toronto-1.6380211.

22   Kari Raymer Bishop, interview with author, Zoom, March 9, 2023.

23   Warnick, *This Is Where You Belong*, 22–23.

24   Alex Cyr, Anthony Milton and Mathew Silver, "The Homecoming Club," *Toronto Life*, October 25, 2022, https://torontolife.com/real-estate/the-homecoming-club-five-families-on-leaving-toronto-regretting-it-and-finding-their-way-home/.

25   Trey Williams, "The Great Pandemic Migration Has Resulted in a Lot of Regrets and Tears," *Fortune*, March 4, 2023, https://fortune.com/2023/03/04/most-americans-have-moving-regrets-pandemic-migration/.

26   Keith Gessen, "King Germ," in *Raising Raffi: The First Five Years* (Viking, 2022), Kobo.

THE REAL CRISIS OF TRUTH

In 2016, "post-truth" was the Oxford Dictionaries word of the year.[1] It refers to a condition in which people's feelings and opinions matter more than objective facts. It doesn't matter that photographs show much smaller crowds at Trump's inauguration than Obama's. Trump says his was bigger, and you believe what Trump says because he's your man (and you hate Obama). You're encouraged by the premier of Alberta saying she'll investigate whether the Pentagon is spraying chemtrails in the skies above her province. You felt safer taking horse dewormer than a Pfizer vaccine to guard against Covid-19. Anti-vaxxers, climate change deniers, 9/11 truthers, teens on TikTok saying Helen Keller didn't exist. All of us lying about our lives, making ourselves seem happier, fitter, more attractive, more depressed or more in love than we actually are, every time we post about ourselves on social media.

If you're looking for an image of post-truth in action, you could do worse than recalling swarms of QAnon faithful attacking the US Capitol on January 6, 2021, in defence of a president who, by one tally, told more than nine thousand lies during his first term in office.[2] QAnon, remember, in the words of *BBC News*, is "a wide-ranging, completely unfounded theory that says that President Trump is waging a secret war against elite Satan-worshipping paedophiles in government, business and the media."[3] A deep-state operative known as Q offers cryptic clues ("Q drops") about the struggle between elites (some of whom are lizard people) and the saviour, Trump. Believers unravel these clues, debating and sharing mind-bogglingly complex interpretations online. Marjorie Taylor Greene, a vocal QAnon supporter, has won a Georgia seat in the US Congress in all three elections since 2020. Trump 2.0's head of the FBI agrees with "a lot of what that [QAnon] movement says."[4]

I once read somewhere that telling a lie is an important milestone in childhood development. Lying requires knowing that people with different experiences might understand the world differently. Fine – I'm at peace with the fact that my kids lie sometimes. But I want Gus and Winnie to value truth, both as a principle to guide living among others and for being at one with themselves. Attempting to find that true self, the authentic core to our being, is part of living the healthiest, most meaningful, individual and social life possible. It's not honesty as a form of discipline, rule-following or commitment to an abstract ethic. It's about thinking and acting in ways that establish the greatest alignment between the reality of the world and the understanding of the self. In his essay "On Liars," Montaigne calls lying "an accursed vice. It is only our words which bind us together and make us human."[5] I am fascinated by, and concerned about, the notion that we've entered an age of post-truth.

The sense that truth was in crisis goes back at least to Stephen Colbert's neologism "truthiness," coined in 2003 during coverage of the US invasion of Iraq.[6] In 2004, Oxford Dictionaries defined "truthiness" as "the quality of seeming or being felt to be true, even if not necessarily true." In 2016, Oxford Dictionaries declared that "*post-truth* extends that notion [of truthiness] from an isolated quality of particular assertions to a general characteristic of our age."[7] What's different in the Trump era, so the story goes, is that truthiness has spread throughout all of society, an idea explored in greater depth in books such as Michiko Kakutani's *The Death of Truth: Notes on Falsehood in the Age of Trump* and Ken Wilber's *Trump and a Post-Truth World.*[8] In Claire Dederer's words: "Where once there were agreed-upon narratives, now there are competing stories, falsehoods and deceptions."[9]

If we're living in a post-truth society, it's going to be much harder, if not impossible, to achieve the kind of alignment with reality that I want for myself and my children. It means that the influential people in my kids' lives – teachers, coaches, friends, the parents of friends, roommates, bosses – will be more likely to dismiss facts in favour of feelings in all they do. They could be anti-vaxxers, climate change deniers, controlled by artificial intelligence, ideologues, dupes to this or that fad. Socially, it means curriculum trending more toward individual tastes than scientific fact; political parties no longer maintaining even the slimmest connection between election platform and approach to governance. The more that truth is degraded across society and the value of truth-telling diminishes, the more difficult it will become for my kids, for anyone, to pursue an existence achieving harmony between what is real and how we interpret reality.

The thing is, I'm not so sure we are living in a post-truth society. I recognize the trends labelled "post-truth," and I'm concerned about the dangers of some of them. For example, leaders of the so-called "Freedom Convoy" in Ottawa in 2022 denied the truth of Covid-19 science and the reality of broad public support for Covid restrictions. They also spewed racist lies about immigration policy and physically threatened striking workers.[10] Instances of right-wing post-truth violence and racism are real and need to be challenged. I'm just not convinced that post-truth trends are dominant across society or that they're notably new.

If we roughly compare majority vs. minority behaviour, look at public opinion polls and so on, we see that vaccination rates against Covid-19 were above 80 percent among eligible populations in Canada and the US by fall 2021.[11] Popular support for pandemic restrictions rose and ebbed but even into the third year of the pandemic, they often reached higher than 70 percent.[12] That's not a society of anti-vaxxers.

Climate change deniers are a fringe group. There's no consensus about how to respond to the ecological crisis, and the leading policy options are woefully inadequate. But this is standard democratic disagreement, not a new form of mass truth-denial.

Communication scholars proclaim that developments in artificial intelligence have "made 'truth' an endangered species and also fostered a strategic deployment of 'lies.'"[13] They point to the case of Cambridge Analytica, the British consulting firm that manipulated elections by harvesting massive amounts of online personal data without the consent of users to create targeted political advertisements. Researchers in Japan explain that AI-based systems increasingly control not only people's access to information about politics and society but also people's self-conception through "pseudo-personalized data services," such as social media feeds and online ads.[14] A leading professor of psychology and neuroscience puts it bluntly: "Reality has become pixels, and pixels are now infinitely inventable [. . .] We can create them any way we want to."[15]

Yet different experts argue no less enthusiastically that AI-technology is our best hope for avoiding a post-truth future. In the *Bulletin of the Atomic Scientists*, John Cook expresses confidence in the war against fake news because of the ability of AI to "automatically detect a claim in real-time and instantly assess its accuracy."[16] In 2024, Cook and his colleagues shared results from their "fallacy detection" machine-learning model designed to debunk misinformation on climate change.[17] The model is part of the growing field of "technocognition," which integrates research from psychology and computer science in search of "the holy grail of fact-checking."[18] Technocognitivists believe they are on the cusp of creating AI models that reveal facts at the same time as nudging users to become more appreciative of truth, even when truth contradicts users' interpretive frameworks.

Existing AI tools, such as the know-it-all generative AI machine ChatGPT, are technically impressive, and disrupt certain professions (especially in academia, journalism and the arts). But, according to the sci-fi novelist and journalist Cory Doctorow, these tools are "fun and playful" experiments, not the stuff of an economy-altering paradigm shift.[19] In truth, says Doctorow, AI is rather bad at even basic tasks, such as drafting tax returns or reading X-rays. Machine-learning models often "'hallucinate' and confabulate," requiring "a 'human in the loop' to carefully review their judgments." Needing more human oversight is the opposite of what true believers (and cutting-edge tech business proposals) promise AI will do. Doctorow does the math on the claims of AI true believers, asking what return-on-investment could you realistically expect from a tool that can't even be trusted to file your taxes, let alone drive a car, harvest crops, drill for oil, win in court or wage war.

> Add up all the money that users with low-stakes/fault-tolerant applications are willing to pay [that's you and me and my students messing around on ChatGPT]; combine it with all the money that risk-tolerant, high-stakes users are willing to spend [this is venture capital hunting for the next big profit-maker]; add in all the money that high-stakes users who are willing to make their products *more* expensive in order to keep them running are willing to spend [that's a niche business model, luxury brand stuff]. If that all sums up to less than it takes to keep the servers running, to acquire, clean and label new data, and to process it into new models, then that's it for the commercial Big AI sector.

Doctorow's argument is that there simply isn't the profit-incentive to spend what would need to be spent for AI to become either the engine of, or saviour from, a radically new era of post-truth. We're not on the cusp of a technological revolution; we're on a speculative bubble certain to burst sooner or later. The question we should be asking ourselves, says Doctorow, is this: What will the bubble leave behind after it pops? When venture capital flees the commercial AI sector, what tools, networks and knowledge built while the bubble grew might we salvage and redirect toward new experiments in democratization and satisfying human needs, maybe even to elevating truth?

What does AI itself have to say about its relationship to post-truth society? On a grey, insufferably humid morning in July 2024, while drinking more coffee than recommended by any AI-nutritionist worth its synthetic salt, I typed into ChatGPT's query box: "Is A.I. leading to a post-truth world?" Here's how our robot overlords responded:

> The impact of A.I. on truth and information is a complex and multi-faceted issue. On one hand, A.I. technologies can be used to spread misinformation and manipulate public opinion through techniques like deepfakes and automated disinformation campaigns. These can contribute to a sense of a post-truth environment where facts and objective reality can be obscured or distorted.
>
> However, A.I. also has the potential to enhance truth-seeking and fact-checking efforts. Natural language processing (NLP) models can be used to analyze vast amounts of information, detect patterns of misinformation, and assist in verifying the accuracy of claims. A.I.-driven tools are being developed to combat fake news and enhance the credibility of information online.

I'd say that's actually a pretty good answer, somewhere in the B- to B range. (No deeper analysis in relation to broader social dynamics? No A from me.) Unlike the technophobic line, ChatGPT's answer frames technologies as tools, created by humans, capable of multiple, varied uses. The question of how AI will affect the status of truth is social, political. It will be answered not by source code, machines, irrational forces but through struggles among individuals and groups, all of whom have the capacity for truth-telling, lying, critical rationality and magical thinking. To interpret today's technological advances apart from similarly disruptive inventions in the past (think steel, the printing press, electricity, air travel, telephony), inventions which no less radically challenged socio-economic relationships and commonsense notions of reality, is to not tell the whole truths of history or present. Dare I say it fibs more than a little about the uniqueness of truth being attacked in our time?

Today, more people than at any point in history have completed some form of post-secondary education. This explosion of mass higher education means more young people than ever are spending a few of their formative years in institutions that value and teach scientific methods of truth-seeking. Steven Shapin, an historian of science, says that what's more remarkable than pockets of science skepticism among the population is that "large sections of the public have encountered and, without friction or frisson, accepted a mass of scientific claims – facts and inferences from facts."[20]

If you think about the most important things you do in a day, and the vast majority of the people you interact with while doing those things, I'll bet that your experiences are most often, and most meaningfully, shaped by common standards of truth, and widely-shared agreement about the importance of truth. Play yesterday back in your mind. Here's my wager: virtually every step of the way, you told the truth, and others told the truth to you. At your job, on the bus, at the grocery store, around the dinner table. And I'll bet that when you encountered something untruthful – an exaggerated claim in a TV advertisement, a friend's Facebook post about how great their trip to the apple orchard was – your internal bullshit detector went off, and you either dismissed the information completely, or accepted it only partially, knowing it wasn't the full truth.

Complicating matters further, many people who we might think of as exemplifying post-truth sensibilities consider themselves to be proponents of truth. I think of the professor on my campus who required his class to write an essay opposing vaccine science. This anti-vax colleague of mine believes in the importance of truth. In fact, he claims to be the real truth-teller, as opposed to the rest of us duped by Big Pharma and Orwellian government. The guy is wrong. Vaccine science is clear. But he's not against truth. He's for truth. (At least, that's what he yells when showing up to faculty meetings every few years.) The vast majority of professors, whether they hit on truth or not, are devoted to truth as defined by philosopher Harry Frankfurt: that is, that which corresponds to reality.[21]

When exceptions appear, they can fascinate us. I once had a student ask whether I agreed with her that QAnon is the contemporary manifestation of the Enlightenment. "QAnon is saying exactly what Kant said," my student told me during office hours. "Think for yourself, don't trust the authorities, things are not always what they seem." Use your powers of critical thinking to find out the truth – don't expect the government and the church to hand it to you. Only the skeptical are free. You have to dig deeper, do your own research, if you're going to see through the lies of society.

In the middle of our conversation, I realized something I should've known going in: It's not that followers of QAnon don't believe in truth. Rather, my student believes that QAnon is telling the truth; it's the rest of the world that's lying or refusing to see what is true.

I told the student her interest in QAnon seemed to me to reflect reasonable skepticism in official representations of politics and economy. Political and business elites tell us that governments need to cut because there's not enough money to go around, yet we see the top layer of society living lives of untold wealth. We're told we live in a democracy, yet have no experience of meaningful participation in major decisions, and things go on more or less the same no matter which party is in power. Official stories don't correspond to reality. Here's a group, QAnon, saying exactly that, and offering an alternative explanation that cuts through the surface-level bullshit.

The problem with QAnon, I said, is that it forfeits the best of our methods for arriving at trustworthy knowledge: namely, the deliberate movement between empirical research and formal theorizing. Without even questioning QAnon's conclusions (and as much as I never liked Joe Biden, I doubt that he is a lizard), I don't trust sweeping claims about society that haven't been developed through a combination of systematic observation and social theory. The point of Kant, Diderot and other champions of Enlightenment wasn't that anything goes – your opinion was no better or worse than any other. It was that we can get closer to the truth by exercising our capacity for critical judgment, rationality, weighing competing explanations and eliminating those that fail to stand up to rigorous scrutiny.

"Exactly," said my student. "That's exactly what Q is doing. He's eliminating the false explanations, the ones that don't hold water. If we think critically about the connections he's pointing out, then we'll see the truth."

I left the conversation disappointed in myself for not disabusing her of her attachment to QAnon. Oddly, then, I also left much more sympathetic to followers of QAnon than I'd been before. But I also left convinced that the pursuit of truth is about methods of inquiry, not the authority of institutions or individuals.

In *Post-Truth: Knowledge as a Power Game*, Steve Fuller argues that Western societies have always been post-truth. In fact, says Fuller, the fight isn't really between pro- and anti-truth forces. It's a fight over the conditions under which truth can be decided. Fuller says the fight goes back to Plato and the Sophists, "post-truth merchants, concerned more with the mix of chance and skill in the construction of truth than with the truth as such."[22] Plato's *Republic* envisions the best society as the one in which philosopher kings rule on the basis of the noble lie (i.e., that at birth individuals are imbued with different metals corresponding to different social roles). The history of Western philosophy and statecraft, says Fuller, is rooted in Plato's post-truth campaign, and continues to manipulate the masses through deception. For this reason, Fuller argues we need *more* post-truth – more radical attacks on the reigning truth-regime – not less. There is a democratic force in post-truth skepticism.

I agree with Fuller in certain cases. There certainly were anti-elite, democratic currents in rejections of pro-EU referenda in Greece (2015) and the UK (2016). Even Trump's election in 2016 involved elements promoting deeper democracy. (Hillary Clinton was a breakthrough candidate in terms of gender; but compared to Trump, she was the candidate of politics-as-usual.) But Fuller's provocation fails to address the fact that the leading force of what we call post-truth today comes from a hard-right that substitutes true belief for genuine skepticism and often encourages political violence to achieve racist, xenophobic, sexist, anti-queer ends. Where I lived recently, in Hamilton, Ontario, arguing for "post-truth" in the abstract means siding with groups who deny systemic racism is a problem in schools and who depict the cops as the innocent good guys. By contrast, the banner of truth is carried by social justice organizations such as the Hamilton Centre for Civic Inclusion, the Hamilton Encampment Support Network and Stop the Spread – groups highlighting the truth of systemic inequality and demanding structural change.

In Naomi Klein's award-winning *Doppelganger: A Trip into the Mirror World*, the power of post-truth politics lies not, or at least not primarily, in the charm, or media savviness, or depravity of far-right figures such as Donald Trump, Steve Bannon, leaders of the Freedom Convoy or (Klein's evil doppelganger) Naomi Wolf. Nor is the problem that millions of people are stupid or hardwired for racism. Understanding the appeal of post-truth politics requires recognizing that there is "some truth mixed in with the lies" of even laughably distorted conspiracy theories.[23]

It is true, for instance, that the governing parties of the post-1970s neoliberal status quo, whether of the right, centre or left, have overseen "devastating collective failure," such as growing inequality, catastrophic ecological damage and unending war, for which they deny responsibility and to which they redouble commitment.[24] This absence of honesty opens space for, and doubtless encourages, deceitful and harmful explanations for systemic injustice and misery. The increase in people's susceptibility to post-truth interpretations is directly tied to the increasingly hypocritical, hollow politics of neoliberal times. According to Klein, MAGA didn't birth post-truth society; rather, post-truth is an outgrowth of a politically diverse range of mainstream institutions that have espoused the principles of democracy, community, inclusivity and compassion, while enacting policies and acting in ways that are anti-democratic, unfair, inhumane.

Klein is especially damning of government rhetoric about the importance of community and national health amid the gutting of welfare state social programs. Throughout the neoliberal period, while tuition fees and rent prices shoot up and welfare supports are slashed, we've seen "every hardship and every difficulty – from poverty to student debt to home eviction to drug addiction – pathologized as a personal failing," while every success is treated "as proof of the relative superiority of the supposedly self-made."[25] The facts of social interdependence and the structures of social power that significantly determine individual success or failure are ignored. Meanwhile, billionaires are celebrated for creating jobs while they bust unions and pay workers less than a living wage. Governments apologize for historical wrongs against Indigenous people yet continue to underfund Indigenous communities and exploit Indigenous lands. Self-identified progressive politicians crush legal strikes through back-to-work legislation. Government and business leaders make urgent announcements about the need to address the ecological crisis while administering systems guaranteed to worsen the crisis. You might say that post-truth in politics, rather than being the territory of the far-right political fringe, describes politics-as-usual.

As a result, aside from the purest of Pollyannas and party dupes, hardly anyone genuinely believes the rhetoric of political and economic elites anymore. We're not facing a culture of "post-truth" but a bunch of lies that virtually everyone sees through. Amid the hellscape of official politics today, who doesn't realize we've been sold a bill of goods? According to the Ipsos Global Trustworthiness Index, politicians are "the world's least trusted profession."[26] Little wonder, says Klein, why it's now common sense to reject the official line that modern democracy embodies government for and by the people: "The mechanics of oligarchy are not hidden."[27]

In December 2024, a few weeks after Luigi Mangione was accused of assassinating the CEO of UnitedHealthcare, one of the most profitable health insurance companies in the United States, one poll showed that nearly half of young Americans believed that the killing was somewhat or totally justified. Only 17 percent of those polled expressed sympathy with the dead CEO.[28] While elites expressed shock and disgust at the outpouring of public support for Mangione, ordinary Americans felt otherwise. Millions of people whose lives have been destroyed by the high cost of insurance, by having claims denied or by being excluded from coverage in the first place, they know too well that the system is rigged against them.

In the context of knowing the official story is bunk, it is rational to seek alternative explanations for the state we're in. In Klein's words, while conspiracy theorists often get the facts wrong, sometimes wildly so, they "often get the *feelings* right [. . .] the feeling that every human misery is someone else's profit, the feeling of being exhausted by predation and extraction, the feeling that important truths are being hidden."[29] When political leaders fail to act, or their actions fail to matter, "inevitably, people reach for narratives to make sense of this reality."[30] The problem is not that people are searching for hidden explanations for our perilous situation. The problem is that, in our troubled times, when we face the real threats of climate catastrophe, extreme socio-economic inequality and precarity, the right-wing has far more effectively drawn disaffected minds to its side, even though critical theory in the Marxist tradition has always looked for the truth of human power relationships, exploitation and the potential of democratic movements, below the level of everyday experience. How has the right managed to do this?

It's not as though the far right's stories are more appealing because they're more honest. On the contrary, critical social science and investigative journalism remain the most rigorous methods for producing answers approximating truth about society today. But the right wing is offering easy answers at a time when progressive movements are demobilized and deep social change for the better seems impossible. Klein's "Mirror World," the alternative political reality in which vaccines are bad and demagogues are democratic saviours, has plenty of answers offering people who are suffering and frightened the emotional rewards of certainty and vengeance. When the answers people find are promoted across social media, easy to comprehend, align with existing identities (whether nationalistic, racist or that of the anti-vax "wellness industry" chronicled by Klein[31]), they can be appealingly grounding, reassuring, inspirational.

The cause of your suffering is Chinese communism, or chemtrails, or bloodthirsty feminists, or globalists, or Mexican criminals. These kinds of answers have thrived in the vacuum created by the collapse of the broad left of the twentieth century and the collapse of the broad welfare state that, for several decades after the

Second World War, provided some people some level of economic protection and sense of common political identity. And as the far right builds robust political infrastructure partially through peddling simplistic, emotionally satisfying explanations for causes and solutions to unfairness and suffering, it is assembling surprisingly large, diverse coalitions of groups and individuals. Klein observes that Steve Bannon doesn't need you to be a MAGA purist at the outset. He's merely providing new language, new frameworks for understanding what's gone so wrong with the world, guiding you down the path of disillusionment with the norm, confident that in today's political context, you'll find your way into MAGA's big tent. What, then, to do about the appeal of conspiracy theories, far-right pundits and policy proposals, and the growing receptivity to them?

The mainstream critique of post-truth society yearns for the restoration of a pre-Trump politics. Books such as Lee McIntyre's *Post-Truth* and Sophia Rosenfeld's *Democracy and Truth* mourn paradise lost. Sure, people have always lied, marketing exaggerates and governments cover stuff up. But, the anti-post-truth literature suggests, there was a time, not so long ago, that everyone knew the difference between truth, partisan allegiances and gut feelings. Social scientists warn of the shift to a post-truth landscape, where "the normative foundations of democracy and freedom, namely reason and truth," are shattered, "leaving societies in a perpetual state of crisis."[32]

If only we could straighten out Trump's lies and the erratic politics they produce. If only people learned to be more savvy social media users. If only colleges and high schools did a better job teaching information literacy. If only we could end the distractions of the past thirty years of gotcha politics, we could reorganize society around the cardinal virtue of truthfulness. We could return to a society prior to post-truth. How to get back there?

For McIntyre, the answer comes down to cultivating the determination of honest, clear-thinking individuals resolving "not to adjust to living in a world in which facts do not matter, but instead to stand up for the notion of truth and learn how to fight back."[33] His demand that "one must always fight back against lies" is at odds with his book's core argument that facts fail to sway people largely because of cognitive biases. His assurance that "facts about reality can only be denied for so long" rings hollow after telling readers that millions of people, because of ideological blinders, fail to accept what's plainly true. McIntyre counsels not to yell at someone when you're trying to convince them of the facts. Be patient; be civil. If you can, show them a graph. Self-criticism, too, is important, because "whether we are liberals or conservatives, we are all prone to the sorts of cognitive biases that can lead to post-truth." Nothing more substantive is offered beyond: check your own dubious assumptions about the truth; support better media outlets; speak up.

Even Rosenfeld, whose analysis is far more sophisticated, and who traces struggles over truth in politics back to the 1700s, ends her book with practical suggestions for reinforcing an Enlightenment-style truth regime. She cheers journalists exposing the lies of the powerful, stresses the importance of maintaining the integrity of elections and an independent judiciary, calls for social media giants to be turned into public trusts, suggests tightening free speech laws.[34] Picture pre–George W. Bush liberal democracy with slightly stricter rules around who can say what. This is the mainstream anti-post-truth side's vision of how to resolve the truth crisis.

The vision includes some reasonable policy proposals. We can debate what chance they have of being adopted. It seems to me, however, badly mistaken to assume that wish lists like these, even if magically granted, would take us back or move us ahead to a social order rooted in truth.

The reason that the mainstream critique of post-truth is inadequate is that it fails to recognize the necessity of untruth, deceit, misrepresentation, in sustaining the pre-Trump truth regime it wants to recover. Mainstream critiques of post-truth ignore the fact that capitalism and liberal democracy, even when working perfectly according to their own logic, depend on world-defining myths that cover up truths of historical and present-day human connectedness. I'm not talking about myths like being brainwashed under the rule of the lizard people. I'm talking about the indispensable role played by untruths in reproducing mass consent for the official political and economic systems of the modern era. I'm saying that, under capitalism and liberal democracy, in stark contrast to their founding claims of promoting freedom and collective self-government, people and the earth are systematically exploited, and popular sovereignty is severely restricted. There is a crisis in truth running through the history of modernity far more pervasive than the one fussed over by contemporary critics of "post-truth."

I'm all for research and public debates figuring out how to combat the spread of digital disinformation, the conspiracy theories feeding today's far right and other expressions of post-truth that are unique to the Trump era. But as long as we're talking about the value of truth and the poison of untruth in politics, let's use this moment to think more expansively about the distortions of modernity's founding myths. These myths have repressed the true sources of human suffering, the true possibilities of social life for hundreds of years before Trump stepped onto the scene. Let our dreams of a *post*-post-truth era of greater fidelity to reality, fuller honesty in politics, not be limited to resetting the clock to 2015, but instead be geared toward unleashing a future that reckons with the myths upholding modernity's defining features.

Consider the way that, for centuries, capitalism has been

framed as a driver, defender and expression of freedom. In the eighteenth century, Adam Smith wrote that under capitalism, "every man . . . is left perfectly free to pursue his own interest in his own way."[35] Free citizens exchanging goods and services on the free market. Your choice how to make and spend money. Milton Friedman, the famed University of Chicago economist, said capitalist economic arrangements "play a dual role in the promotion of a free society. On the one hand, freedom in economic arrangements is itself a component of freedom broadly understood, so economic freedom is an end in itself. In the second place, economic freedom is also an indispensable means toward the achievement of political freedom."[36] Since the end of the Second World War, one of the principles of US foreign policy has been spreading free markets in the name of spreading political freedom.

Are we truly free under capitalism? You could stop showing up to your job. But unless you're one of the very few people with endless piles of cash squirreled away, or you're able to live secretly and self-sufficiently in the woods, you're going to need money to survive. Technically, it's true: you don't *need* to work. No one has a gun to your head; you're free to do whatever you like. In fact, you definitely *do* need to work.

In *Capital*, Karl Marx's classic critique of the market economy, he referred to "the silent compulsion" of capitalism forcing the vast majority of the population into wage labour.[37] Under capitalism, Marx wrote, most people are "free in the double sense that as free individuals they can dispose of their labour-power as their own commodity, and that, on the other hand, they have no other commodity for sale."[38] Free from legal bondage, and free (read: *prevented*) from accessing the land, housing, food and other resources necessary to live without wages (because those are controlled by major property owners and investors, i.e., capitalists). In contrast to the depiction of capitalism as a world of individual freedom, in truth, all but the extremely wealthy have no choice but to spend most of their waking lives working for a wage.

What's more, there's very little freedom at the workplace itself. No free speech (try talking back to your boss sometime or making impromptu political speeches from behind the cash register), and hardly any control over how the job is done. Many jobs, especially in sectors with high concentrations of women, require forms of emotional labour – smiling, soothing, keeping quiet – that control not only workers' bodies, but their feelings during work hours. The extreme disconnect between workers' wages and CEO earnings puts paid to capitalism's founding myth that all you need is to work hard to succeed. No matter how hard the average person works, they're never going to be a billionaire. By ignoring the unfreedom of capitalism, indeed, by treating the status quo ante as the proper place to return to and rebuild a world of truth upon, the critics of "post-truth" society help reproduce and cover up capitalism's lies.

As for liberal democracy, it is doubtless a version of rule by the people. Regular elections to choose government combined with a slate of civil rights remain the basis of political rule throughout Canada and the Global North. The first Trump administration was awful, and attempted to weaken democratic institutions; but it wasn't fascist (we'll have to see about Trump 2.0, I guess), and didn't crush countervailing forces it claimed to hate so much (i.e., Congress, the press, the courts and the electoral system). Nevertheless, the extent to which liberal democracy truly expresses popular sovereignty is routinely misrepresented by its proponents. (And I've chosen the verb with care: "misrepresent," says *Merriam-Webster*, is "to give a false or misleading representation of.")

Citizens send representatives to legislatures, and elected Members of Parliament pass laws on behalf of the people. However, in liberal democracy, the basic laws of capitalism ensure that major economic decisions are beyond the reach of popular power. Even if the will of the community is virtually unanimous and obvious, if it contradicts the will of major property owners in the region, expressions of democracy are virtually meaningless. For example, no one in Leamington, Ontario, wanted the Heinz plant closed in 2014. Heinz's decision to throw 740 people directly out of work (to say nothing of the thousands more who lost jobs in secondary industries) gutted this community of 27,000. In 2018, no one in Oshawa wanted the General Motors (GM) plant closed. But Heinz and GM left anyway, wrecking the lives of countless people, because economic decision-making power lies in corporate hands, not in local communities. Liberal democracy cordons off large pieces of the economy from the meddlesomeness of popular power.

In Canada, majority governments are almost always elected by less than a majority of the voting population, notwithstanding talk of majority-rule. In parliamentary democracies like Canada's, the executive must control the legislature, meaning governance far more resembles rule-by-fiat than negotiation, deliberation and compromise. And I've said nothing about unequal access to policy-makers when comparing business elites and highly-paid lobbyists to community groups and marginalized people.

Considering the wide gap between lofty talk of rule by the people and the actual undemocratic relationships organizing so much of our experiences, is it any wonder public trust in governments is so low, and that voter rates have fallen in recent decades?

In Canada, the US and other nations, the very claim to national sovereignty requires denying the truth of Indigenous sovereignty prior to the expansion of European colonialism. All the scholarship I've read that condemns the "post-truth" of Trump treats the truth-denying fact of settler colonialism not only as unremarkable, a given, but as the even playing field upon which a return to the truth in the future may be pursued. As Lee Maracle writes in *My Conversations with Canadians*, the image of Canada as a legitimate political institution built upon Enlightenment principles of democracy, freedom and rights contradicts reality. It ignores the truth about stolen lands, genocide, treaty promises, the continuing dispossession of Indigenous peoples and their relationship to Turtle Island.

The anti-post-truth literature has nothing to say about the myths of modernity: the architecture of deceit that undergirds Western societies. Considering the insidiousness and global influence of such myths, we shouldn't be surprised to see symptoms of this much more pervasive crisis of (social, historical) truth in the form of conspiracy theories, fake news, fabricated online personalities and moral panic over artificial intelligence. By so restricting our focus on what lies are worth worrying about, and which institutions are best equipped to protect the truth, the anti-post-truth literature distracts us from far more fundamental untruths than any of Trump's whoppers or the yarn QAnon spins. The best critiques of new problems of post-truth will refuse to be complicit in much deeper distortions of reality.

Movements like Black Lives Matter, Idle No More, the Global Climate Strike, and the campaign for Boycott, Divestment and Sanctions against Israeli apartheid have ensured that the myths of modernity are discussed on a broader stage more regularly than at any point in recent decades. Acknowledgement of this more deep-rooted crisis of truth has even found expression in some mainstream political parties, such as Jeremy Corbyn's Labour Party and Bernie Sanders' wing of the Democratic Party. Where concern about "post-truth" ends up – on the side of structural critique or in the defence of status quo capitalist democracy – remains to be seen.

Helping my kids defend against the sharp edge of "post-truth," as typically defined, seems straightforward enough, which isn't to suggest that it's easy. Over dinner, and when we read together, we talk about the importance of truth as a concept. My partner and I model honesty in our interactions with the kids and each other.

We invite questions about why we trust certain ideas and not others. Like a good member of the chattering classes, I've even used tips I read in the *New York Times* for talking to my kids about AI chatbots.[39]

The much more difficult challenge is to resist being complicit in the myths of modernity. The threat of post-truth is more restricted and easier to identify than the crisis of truth going to the core of liberal democracy. For example, despite my anti-nationalism, I've taken Gus and Winnie several times to the Canada Day parade in Port Dover, a small town on Lake Erie. We have family in the area, and it's fun to sit in the shade of a tree on the quaint town's main street while watching the small floats pass by. A team of young hockey players waving from the back of a flatbed truck. Clowns driving old tractors. Fire trucks with lights blazing. I don't wave a flag, though just about everyone else does. Red, white and maple leaves: on faces, on T-shirts, hanging from front porches. Not a festival I identify with politically, but a pleasant day for gathering with friends and family before splashing in the water at the beach.

But think of what an event like this does when it comes to spreading disinformation and hiding truths. Nobody sits my kids down and says, "Canada is a peaceful, happy, safe country. Everyone is included in our community, and for that we should celebrate." Nobody has to say it: That's the message in every maple leaf painted on a kid's face, in the fireworks exploding above the huge crowd at the fairgrounds at night. The truth of Indigenous dispossession, the truth of economic inequality among Canadians, the truth of corporate Canada's pillaging of Third World countries, the truth of the violence against marginalized peoples routinely meted out by Canada's police forces – all of this is hidden. It's not just that these truths aren't on display; rather, they are actively denied by the symbols, atmosphere and implicit claims of the ritual.

Compare July 1 in Port Dover to the hours Gus and I stood with demonstrators occupying the forecourt of Hamilton City Hall in winter 2020. Homeless people and their allies built a small tent city on City Hall's doorstep to demand better funding for shelters and more socially-just housing policies. We brought diesel for the encampment's generator. Activists gave us a tour of the site. Gus asked why our friends had to sleep outside in the cold. The occupation made clear the truth of unequal access to housing, and the truth of the potential of community power-from-below. In blockades – at Hamilton City Hall, at 1492 Land Back Lane, on railroads and over oil pipelines – the truth of history is closest to the surface, becomes easier to access, harder to avoid.

The real crisis of truth can't be addressed solely at the intellectual or behavioural level. It's not just about doubling down on the importance of fact-telling, nor about refusing to participate in post-truth politics. "What needs to take place at the current conjuncture," according to Johan Farkas and Jannick Schou in a rare scholarly critique of the idea of post-truth, "is not the cancellation or shutting down of political conflict, either by policing what is deemed as false or algorithmically sorting the voices who have access to the public sphere. Instead, we need genuine political alternatives and open discussion."[40] I agree. But I'd find their argument more compelling if it grappled with the fact that changing how people think, which, surely is the aim of anti-post-truth politics, is inseparable from taking physical action, from attending to what Alan Sears calls "the erotic dimensions of active resistance." Erotic in Sears's sense cannot be reduced to sexual pleasure; rather, it's his term for understanding politics as the terrain of the whole person, as opposed to an activity exclusive to the mind or the well-delivered speech. As Sears writes, our ideas and feelings take shape through interacting with others: "We develop intense connections as we put our bodies on the line and process our experiences of agony and ecstasy in struggle."[41]

Building the "genuine political alternatives" envisioned by Farkas and Schou, creating the context for open discussion, these are fundamentally embodied projects, dependent on words and symbols, of course, but capable of development only through workplace organizing, strikes, mass demonstrations, occupations and other forms of what the sociologist Frances Fox Piven calls "disruptive politics."[42] Embodied disruption is essential to repairing the rift between word and deed that epitomizes the post-truth politics of both the far-right and mainstream liberals. Naomi Klein writes that "the outright denialism of the [far-right] Mirror World is made thinkable by the baseline war on words and meaning in more liberal parts of our culture."[43] She means that saying one thing while doing something different has become standard political fare. The embodied disruptions of democratic social movements reject this war on words and meaning. In the service of expressed political ideals, embodied action demonstrates the truth of what's been said. It can, while exposing myths of modernity, redeem truth in the unity of word and deed.

In the final pages of *Doppelganger*, Klein suggests that taking collective action is a potent antidote to post-truth politics because when people organize together they are more likely to experience the reality of shared circumstances and interests: "Stuck in the realm of words, we will never run out of reasons to fracture. But when we take action to change material circumstances – whether trying to unionize our workplaces, or halt evictions, or free political prisoners, or build alternatives to policing, or stop a pipeline, or get an insurgent candidate elected – those tensions do not disappear, but they are often balanced by the recognition of shared interests, the pleasures of camaraderie, and, occasionally, the thrill of victory."[44]

Critiques of post-truth society, virtually without exception, make it seem as though our problems will be solved primarily at the level of ideas. They fail to grasp that a genuine politics of truth must be far more dynamic and radical, seeking to transform the foundations of social, political and economic life, so that the ways in which we represent ourselves – as free, as democratic, as committed to environmental sustainability – align with the aims and capabilities of the actual institutions organizing society.

Transcending the real crisis of truth is, therefore, a social process requiring collective action. It can't be done by any one of us alone, nor by a smattering of progressives modelling radical honesty. Nevertheless, one precondition for resolving the crisis of truth is recognizing and rejecting complicity in the myths of modernity. It's just as important to figure out how to do that better with our kids as it is to develop media literacy campaigns and warn people away from QAnon.

NOTES

1   "Word of the Year 2016," *Oxford Languages*, no date, https://languages .oup.com/word-of-the-year/2016/.

2   Daniel Dale, "Dale: Reflections on Four Weird Years Fact Checking Every Word from Donald Trump," Facts First, *CNN Politics*, January 19, 2021, https://www.cnn.com/2021/01/19/politics/fact-check-daniel -dale-reflections-fact-checking-trump/index.html.

3   Mike Wendling, "QAnon: What Is It and Where Did It Come From?," *BBC News*, January 6, 2021, https://www.bbc.com/news/53498434.

4   Kash Patel, quoted in Devlin Barrett, "Trump's F.B.I. Pick Sees 'Deep State' Plotters in Government, and Some Good in QAnon," *New York Times*, January 23, 2025, https://www.nytimes.com/2025/01/23/us /politics/kash-patel-conspiracy-theories-qanon.html.

5   Michel de Montaigne, "On Liars," in *The Complete Essays*, trans. M.A. Screech (1588; Penguin, 2003), 35.

6   "Colbert Report Writers – Truthiness and Pun Journals," moderated by Zachary Kanin, *The Paley Center for Media*, YouTube video, October 24, 2011, https://youtu.be/WvnHf3MQtAk ?si=psOwuYKtQskK9y6g.

7   "Word of the Year 2016."

8   Michiko Kakutani, *The Death of Truth: Notes on Falsehood in the Age of Trump* (Tim Duggan Books, 2018); Ken Wilber, *Trump and a Post- Truth World* (Shambhala, 2017).

9   Claire Dederer, "When Truth No Longer Counts, How Does a Memoirist Tell Her Story," review of *To Name the Bigger Lie*, by Sarah Viren, *New York Times*, June 12, 2023, https://www.nytimes.com/2023/06/11/books/review/to-name-the-bigger-lie-sarah-viren.html.

10 Emily Leedham, "Canada's 'Freedom Convoy' Is a Front for a Right-Wing, Anti-Worker Agenda," *Jacobin*, February 5, 2022, https://jacobin.com/2022/02/canada-freedom-convoy-conservative-right-wing-anti-worker-anti-vaccine.

11 Adam Miller, "Why Mandatory COVID-19 Vaccines for Health-Care Workers Could Help Canada Fight a 4th Wave," *CBC*, July 22, 2021, https://www.cbc.ca/news/health/canada-mandatory-vaccines-covid-19-healthcare-workers-1.6111486; Mychael Schnell, "70% of US Adults Are Fully Vaccinated, 80% Partially: White House," *The Hill*, November 1, 2021, https://thehill.com/policy/healthcare/579479-white-house-70-of-adults-are-fully-vaccinated-80-partially/.

12 See, for example, Clifton van der Linden and Alexander G.W. Beyer, "Majority of Canadians Disagree with 'Freedom Convoy' on Vaccine Mandates and Lockdowns," *Conversation*, February 3, 2022, https://theconversation.com/majority-of-canadians-disagree-with-freedom-convoy-on-vaccine-mandates-and-lockdowns-176323.

13 Shankhadeep Chattopadhyay, "Social Media and the Corrosion of Public Discourse: A Critique on the Rhetoric of Post-Truth," *New Literaria* 2, no. 1 (2021): 165.

14 Tatsuya Yamazaki et al., "Post-Truth Society: The AI-Driven Society Where No One Is Responsible," paper presented at the International Conference on the Ethical and Social Impacts of ICT, University of La Rioja, Logroño, Spain, June 2020, section 1, paragraphs 2 and 3.

15 Michael Graziano, quoted in Thor Benson, "Humans Aren't Mentally Ready for an AI-Saturated 'Post-Truth World,'" *Wired*, June 18, 2023, https://www.wired.com/story/generative-ai-deepfakes-disinformation-psychology/.

16 John Cook, "Technology Helped Fake News. Now Technology Needs to Stop It," *Bulletin of the Atomic Scientists*, November 17, 2017, https://thebulletin.org/2017/11/technology-helped-fake-news-now-technology-needs-to-stop-it/.

17 Francisco Zanartu et al., "Detecting Fallacies in Climate Misinformation: A Technocognitive Approach to Identifying Misleading Argumentation," unpublished paper, arXivLabs, May 14, 2024, 5, https://doi.org/10.48550/arXiv.2405.08254.

18 Zanartu et al., "Detecting Fallacies," 10.

19 Cory Doctorow, "What Kind of Bubble is AI?," *Locus*, December 18, 2023, https://locusmag.com/2023/12/commentary-cory-doctorow-what-kind-of-bubble-is-ai/.

20 Steven Shapin, "Is There a Crisis of Truth?," *Los Angeles Review of Books*, December 2, 2019, https://lareviewofbooks.org/article/is-there-a-crisis-of-truth/.

21 Harry Frankfurt, *On Truth* (Knopf, 2006).

22 Steve Fuller, *Post-Truth: Knowledge as a Power Game* (Anthem, 2018), 29.

23 Naomi Klein, *Doppelganger: A Trip into the Mirror World* (Penguin, 2023), 175.

24 Klein, *Doppelganger.*

25 Klein, *Doppelganger*, 231.

26 Jamie Stinson, "Politicians the Least Trusted Profession, While Doctors the Most Trustworthy," *Ipsos*, October 23, 2023, https://www.ipsos.com/en-ca/politicians-least-trusted-profession-while-doctors-most-trustworthy.

27 Klein, *Doppelganger*, 240.

28 Erica Pandey, "Exclusive: Young Americans Sympathize More with CEO Shooting Suspect than Victim," *Axios*, January 9, 2025, https://www.axios.com/2025/01/09/luigi-mangione-approval-poll-gen-z.

29 Klein, *Doppelganger*, 242.

30 Klein, *Doppelganger*, 167.

31 Klein, *Doppelganger*, 181.

32 J. Farkas and J. Schou, "Post-Truth Discourses and Their Limits: A Democratic Crisis?," in *Disinformation and Digital Media as a Challenge for Democracy*, ed. G. Terzis, D. Kloza, E. Kużelewska and D. Trottier (Intersentia European Integration and Democracy Series, 2020), preprint version, 1, https://mau.diva-portal.org/smash/get/diva2:1451301/FULLTEXT01.pdf.

33 Lee McIntyre, "Fighting Post-Truth," chap. 7 in *Post-Truth* (MIT Press, 2018), Kobo. All quotes from McIntyre are from this chapter.

34 See Sophia Rosenfeld, "Democracy in an Age of Lies," chap. 4 in *Democracy and Truth: A Short History* (University of Pennsylvania Press, 2019), Kobo.

35 Adam Smith, *The Wealth of Nations* (1776; Modern Library, 2000), 651.

36 Milton Friedman, *Capitalism and Freedom*, 40th Anniversary ed. (University of Chicago Press, 2002), 8.

37 Karl Marx, *Capital: A Critique of Political Economy*, vol. 1, trans. Ben Fowkes (1867; Penguin Books 1990), 899.

38 Marx, *Capital*, 272. I believe Marx would appreciate my changing his quote slightly to make his point in non-gendered language.

39 Christina Caron, "The A.I. Chatbots Have Arrived. Time to Talk to Your Kids," *New York Times*, March 22, 2023, https://www.nytimes.com/2023/03/22/well/family/ai-chatgpt-parents-children.html.

40 Farkas and Schou, "Post-Truth Discourses," 15.

41 Alan Sears, "Eros and Revolution," *Midnight Sun*, January 13, 2025, https://www.midnightsunmag.ca/eros-and-revolution/.

42 Frances Fox Piven, "The Nature of Disruptive Power," chap. 2 in *Challenging Authority: How Ordinary People Change America* (Rowman and Littlefield, 2006).

43 Klein, *Doppelganger*, 155.

44 Klein, *Doppelganger*, 331.

THE FALL OF CUDDLING, 2018

On the day my son, Gus, turned five months old, I wrote to him:

> One of my favourite things to do with you is lie on my side facing you while you lie on your side facing me. I cover your face and head and neck with kisses. I kiss your ears, being careful not to make a smacking sound so loud it would hurt you. I kiss your mouth, feeling your little lips as two wet thin lines. I kiss your smooth round forehead and your soft cheeks. I make kissing sounds – mwoa, mwoa, mwoa – in rapid, regular bursts.

In Ben George's intro to the essay collection *The Book of Dads*, he admits being surprised to find that even in the twenty-first century, there is very little good writing about being a father. If his Google searches for books on "fatherhood" didn't come back with the question "Did you mean 'motherhood?,'" they offered him a list of books "purporting to instruct me on how to keep my baby alive until my wife got home and how to remain a mack daddy even in the grip of new fatherhood."[1] The axes along which the meaning of fatherhood are charted measure professional costs, character rewards, family responsibilities and conformity to dominant notions of gender.

In *Pops*, the novelist Michael Chabon's collection of essays on fatherhood, he recalls being told by an early author idol of his that writers lose a book with every child.[2] The moral of the story, to writers and artists everywhere: don't have children. The then-childless Chabon does the math of what his future two or three or four children may cost him. And the present-day Chabon, the father of four children, ultimately concludes, with a surprising lack of sentimentality, that in the end we all die, and no one will remember our books or our children anyway. The Norwegian writer Karl Ove Knausgaard writes about loving his children unconditionally, despite how they bore him, require him to do things that diminish his sense of being a man and stand in the way of work.[3]

In Montaigne's "On the Affection of Fathers for Their Children," arguably the first ever essay to explore fatherly love, he warns against becoming attached to sons and daughters too quickly: "If they show they deserve it, we should cherish them with a truly fatherly love."[4] If they turn out to be ingrateful brats, well, too bad for them: "We should, despite the force of Nature, always yield to reason." Montaigne sees babies as not-yet-human. He is "incapable of finding a place for that emotion which leads people to cuddle new-born infants while they are still without movements of soul or recognizable features of body to make themselves lovable." Montaigne is known for his empathy, playfulness, capacity for love. And he celebrates the fierce love fathers may one day develop for their children. But what have babies done for him lately?

In my prep-for-fatherhood reading, I found nothing remotely focused on the embodied experience of uniquely intimate, daily interaction with a baby.[5] Yet I know my new role in the world best through the weight of Gus in my arms, the pinch of his gums chomping on my fingers, the ache in my back from carrying him, the warmth of his head beneath my chin, the wetness of his slobbery cheeks against mine.

What does it feel like to be a father? It feels like pudgy feet that fit in my palms; my cheek on the bristle of a bald spot and the downy soft fluff growing next to it. When I change Gus's diaper, I squeeze his chunky thighs and run my hand over his belly, moving my fingers between the rolls under his arms and around his neck because I assume that feels nice for him, and I know how much

I love the feeling. Being Gus's dad has brought me into a world of physical intimacy, embodied knowledge, that I hadn't known existed.

Last night I took Gus's hand in mine, even though he was sleeping soundly. I wanted to feel his presence in the dark, wanted him to know I was near and that I love him, even through the fog of sleep. I relaxed, in part because I've learned not to make jerky movements when Gus is sleeping, but also because his silent, motionless touch helps me let go of the worries and jitters of the day. I felt more settled, contented, happier, stronger, safer, with Gus sleeping at my side. This is irrational. In an animalistic sense, his presence makes me more vulnerable – to wolves or bears or vampires. And the fact that Gus was sleeping makes it impossible to interpret his movements or sounds as signs of being happy with me.

I'm reminded of the feeling of safety I felt as a boy when sleeping next to Gus's Aunt Kate. I was too old to need to sleep next to her. What was I . . . nine, ten, eleven even? Kate and I shared a room: my single bed in one corner, next to a bulletin board backed with Looney Tunes wallpaper, the bright blue of Yosemite Sam's pants, the bright yellow of Tweety Bird's bulbous head; Kate's single bed in the other corner below a poster of two cartoon bears and red balloon letters reading, Love Makes Life More Bearable.

A near-teen boy sharing a room with his three-years-younger sister is already a problem in lots of people's eyes. Kate and I were best friends. We loved sharing a room together, wouldn't have wanted it any other way. But at night, room-sharing wasn't enough for me. Alone in bed, I felt exposed, afraid. My loneliness was both immediate and transcendent. I heard tires on gravel and feared burglars. The train whistling in the distance was a ghost. Silence was a sign of aliens landing or the loss of oxygen preceding nuclear holocaust. Vampires tapped the windows. Yet I also felt the crushing ache of being alone in the universe.

Lying next to Kate fixed all of that. We didn't touch. Maybe sometimes our legs or bodies would come into contact, but that was rare, fleeting and unintentional. It wasn't about physical intimacy, just nearness, the sense of being together resolved my nightly existential crises. The minute I was next to Kate, I felt entirely better. Car tires stirred gravel on a cozy trip. The train whistle was the sound of sleep in trees. The musty smell of our grey carpet was a sedative. I was happy, confident, sleepy. Often, I'd wake Kate crawling in to lie next to her, then fall asleep before she would. She was smaller than me, younger than me and we were both children. But her closeness fixed my world. I experience something very similar sleeping next to Gus. Why is that? He's just a baby. He can't protect me.

Gus will remember none of this. He won't remember whether I squeezed his calves during diaper changes. He won't even recall whether I did diaper changes. He won't remember the smell of his apricot shampoo or the sting of baby soap in his eyes. He will recall nothing from this first year's days and nights. The writer Paul Auster refers to his time with his infant son as "a lost world," one that can never be recovered. "The boy will forget everything that has happened to him so far. There will be nothing left but a kind of after-glow, and perhaps not even that. All the thousands of hours that A. [Auster referring to himself in the third person] has spent with him during the first three years of his life, all the millions of words he has spoken to him, the books he has read to him, the meals he has made for him, the tears he has wiped for him – all these things will vanish from the boy's memory forever."[6]

What sort of physical relationship did I have with my parents when I was a baby? Did they kiss me as often as I kiss Gus? Did they pick me up the second I fussed? It seems overly deterministic to assume that the way they touched me (or didn't touch me) significantly shaped the person I am today. But if it didn't, if our physical relationship made no impact on my present-day self, then my physical relationship with Gus is entirely for my benefit. It would mean it would make no difference whether I was touching Gus, my son, or a puppy or a warm wet rag. The significance of our touching would be mine alone to draw and would be self-generated. The alternative, however, is also scary. The possibility that every time I touch Gus, and every way I touch him (or don't touch him) has far-reaching, long-term effects, that I might plant seeds of turmoil in my baby that bloom into the crises of his adulthood.

When Karl Ove Knausgaard's six-volume *My Struggle* was on bestseller lists and critics were raving about his honesty, others pointed out that a similar book written by a woman would not have been considered literature, let alone to be a work of observational genius. Knausgaard writes about making breakfast, walking in fields that lead to the sea, hiding in his room to cry, loving his children and hating parts of doing childcare. The slow, detailed descriptions of his experiences are heralded as literary greatness. The same from a woman would've been called diaristic: the quotidian record of a middle-aged woman's life.

I wonder what my position as a dad in a heterosexual couple does to my reflections on the pleasures of embodied parenting. I suspect my position makes it easier to experience the joy of touching baby as a profound discovery. Mothers are assumed to be expert caregivers from the start.

One of the reasons Anne Lamott's *Operating Instructions: A Journal of My Son's First Year* stands out among new-mom memoirs is Lamott's willingness and talent for describing the anger, resentment, at times repulsion she felt while spending time with her baby. ("I heard him begin to whimper, and I thought, 'Go back to sleep, you little shit.'"[7]) The mom who isn't constantly joyful about her motherly role remains something of a taboo. What I am able to celebrate as a miracle is considered to be part of moms' normal work. *Mom loves to touch her baby* is not breaking news.

What about babies who grow up without moms? What about moms who sleep through baby whimpering, who do not experience "mother's intuition"? Are these children and parents failures from the start? Where is the wonder, the awe, the gratitude for mothers who've *discovered* the pleasures of touching baby?

Before Gus got up this morning, I wrote him a series of questions:

Will I delight less in touching you as you age? The answer can only be yes. There must come a time when I can't spend hours a day stroking you, squeezing you, kissing you. Even now, before you're eight months old, I wonder if the period of fatherhood through touching has peaked and will end before too long. You'll develop other ways of communicating; we'll develop other ways of enjoying time with each other. There is much to gain in the new world of speech and mobility, but I will miss knowing you and your love through all the ways we touch each other.

When will you stop putting your hands in my mouth? At what point will we no longer feel comfortable bathing together? A separation will occur, a judgment rendered, a decision made about a fundamental change in our relationship, and, chances are, neither of us will know it at the time.

What will it mean to our relationship, and to my sense of being your father, your parent, your special person, when I touch you no differently than your friends do? It will be an advance, of course. It will mean that you're developing greater autonomy, assuming fuller control over your body. I want this for you, and there is no other way. But where does this period we're in now go? It will live always inside me, but it will always be a secret from the future-you, no matter how vividly I describe it in the years to come.

NOTES

1   Ben George, introduction to *The Book of Dads: Essays on the Joys, Perils, and Humiliations of Fatherhood*, ed. Ben George (HarperCollins, 2009), xvi.

2   See Michael Chabon, "Introduction: The Opposite of Writing," in *Pops: Fatherhood in Pieces* (HarperCollins, 2018), 3–13.

3   For example, see the famous "RhythmThyme class" scene from *My Struggle*, vol. 2, excerpted here: Karl Ove Knausgaard, "My Struggle," *Bomb*, May 3, 2013, https://bombmagazine.org/articles/my-struggle/.

4   Michel de Montaigne, "On the Affection of Fathers for Their Children," in *The Complete Essays*, trans. M.A. Screech (1588; Penguin, 2003), 435.

5   When my friend and editor Noelle Allen heard this story many years after Gus was born, she remarked: "You should've looked to the poets!" Richard Harrison's *Big Breath of a Wish* (Wolsak & Wynn, 1998), nominated for the Governor General's Award for Poetry, explores embodied discoveries in sound and speech made by his two young children.

6   Paul Auster, *The Invention of Solitude* (Penguin, 1982), 166.

7   Anne Lamott, *Operating Instructions: A Journal of My Son's First Year* (Anchor, 2005), 54.

ON FEARING UNCERTAINTY

PARENTING AMID THE ECO-CRISIS

Fear will not allow us to discount anything, not even something that doesn't exist.
– Javier Marías, *The Infatuations*

Wretched is a mind anxious about the future.
– Seneca, *Letters from a Stoic*

You have never known fear until you have a child.
– Harold in Hanya Yanagihara, *A Little Life*

Because I teach courses touching on environmental justice, and because I'm the father of two young children, it's not unusual for me to be asked how I cope with parenting during the ecological crisis. My answer includes the admission that forecasts of ecological catastrophe produce pictures in my mind of Gus and Winnie, some years from now, scuffing through the dust of the dried-up riverbed near our house in Paris, Ontario. They're searching for food, green shoots or bugs. The fish are long gone, and plants grow only where the water used to run. Turning over a rock, Gus grabs at a roach but his thin hands are slow and shaky. Winnie asks Gus, not for the first time, why their mom and I hadn't prepared them for this. We'd driven a car, warmed the bedrooms with space heaters, flown to Florida one winter. Why hadn't we told them disaster would strike so suddenly? Gus defends his dead parents, says we couldn't have imagined how bad it would get or the swiftness of the collapse. But I'm imagining it right now.

Of course, the scene I'm imagining is absurd, a garish mix of Hollywood movie and parental guilt. The rivers near my home have their troubles, but they're not going to dry up in the next eighty years. (Then again, some models of climate change predict that the Danube, the Rhine and other great rivers will dry up if the world warms four degrees above pre-industrial levels by the year 2100.) The harshest effects of climate change will be uneven: droughts here, floods there, neither in other places. Scientists acknowledge that northern climates, like the one in which I live, will actually benefit from climate change in certain ways in the short term. Rising temperatures in southern Ontario will, for a time, mean improved agricultural output. Even the bleakest projections do not show Gus and Winnie battling for survival on a moonscape.

Yet the existential threat of the climate crisis is real. We live in a world of rising sea levels, desertification, routinized extreme weather, fresh water shortages, ocean acidification, worsening air quality, biodiversity loss, the emergence of antibiotic-resistant diseases, and other climate-related harm. David Wallace-Wells' *The Uninhabitable Earth*, a 2019 *New York Times* bestseller, begins with the ominous line: "It is worse, much worse, than you think."[1]

Wallace-Wells argues that for decades climate scientists and journalists have played down the actual threat of climate change. Experts feared that if people understood how bad the situation is, how far along eco-destruction has already progressed, it would be even harder to turn around destructive trends. By contrast, Wallace-Wells wants to scare people into action by providing a transparent, comprehensive overview of the leading scientific research. This is what Earth will look like when carbon dioxide emissions have warmed average global temperatures two degrees above pre-industrial levels (constant droughts across southern Europe and the Middle East, malaria spreading rapidly, a 60 percent increase in armed conflict, tens of millions of climate refugees). This is what four degrees warmer will look like (constant heat waves, Bangladesh under water, most of Africa a desert, the collapse of grain production, perpetual war). "In a six-degree warmer world," Wallace-Wells wrote in the *New York* magazine article that spawned his book, "the Earth's ecosystem will boil with so many natural disasters that we will just start calling them 'weather': a constant swarm of out-of-control typhoons and tornadoes and floods and droughts, the planet assaulted regularly with climate events that not so long ago destroyed whole civilizations."[2]

In a 2019 interview, Wallace-Wells reiterated that there is no avoiding the effects of significant global warming: "We're not going to get below 2 degrees, and we're on track for something like 4 by the end of the century."[3] In the year 2100, Gus and Winnie will be eighty-two and seventy-nine, respectively. My parents are about that age today. Chief Dana Tizya-Tramm of the Vuntut Gwitchin First Nation (Old Crow) in the Yukon says, "It's going to be the blink of an eye before my great-grandchild is living in a completely different territory, and if that's not an emergency, I don't know what is."[4] Notwithstanding the exaggerations of my eco-apocalyptic nightmare, there is plenty of reason to worry.

When I worry about the climate crisis, I'm in good company. In *Generation Dread: Finding Purpose in an Age of Climate Crisis*, Britt Wray tells stories of how ecological destruction has unleashed "tidal waves of grief, anxiety, pessimism and existential dread" in people all over the world.[5] People living on the frontlines of climate change – those surviving floods in Bangladesh, forest fires in Australia, hurricanes in the Caribbean – may deal with especially severe past trauma and fears of the future. But, Wray explains, even people without direct experience of eco-disaster are increasingly distraught by the state of the planet. Her interview subjects describe panic attacks, insomnia, relationship trouble, suicidal thoughts, all as result of eco-anxiety, what the American Psychological Association defines as the "chronic fear of environmental doom."[6]

Polls suggest that millions of people, young people especially, struggle with eco-anxiety in varying degrees. Catherine Hickman's research describes young people furious with their parents for giving birth to them amid such dire times. Wray herself grappled for years with a question I hear often from my students: "When you confront head-on what scientific models say about the suicidal track we're on, alongside the political establishment's completely inadequate efforts to address it, is it okay to decide to bring a new person into this situation?"[7]

We know what the most obvious expressions of the eco-crisis look like: clear-cut forests, melting permafrost, the ocean on fire, species extinction. Wray wants us also to consider the widespread trouble people have in making sense of the eco-crisis and their lives in relation to it: to think about human-inflicted harm to the non-human environment, and eco-anxiety, as part of a much broader "planetary health crisis."[8] Denying the crisis, disavowing our responsibility to do anything about it, is one dysfunctional response. Then there are those who break down completely in the face of climate apocalypse. How do we reduce the harm of climate change as much as possible is the essential question of our time. Wray says that part of answering that question is addressing another: How do we conceive of the crisis in a way that makes action psychically possible? How do we manage eco-anxiety?

At the start of each semester, I show my classes pictures of my family. Later in the term, when we talk about the climate crisis, students ask about my kids by name: "Are you hopeful for Gus and Winnie's future?" A student once put the question this way: "What hope do you have for Gus and Winnie?" and I've never forgotten the turn of phrase.

I have stock answers. The future is open-ended, I remind them. If mass movements for sustainability surge to dominance in the next decade, the harm caused by climate change could end up being considerably less than what seems likely today. I point them to Meehan Crist's article that frames the question of using Earth's finite resources sustainably away from talk of population limits toward "the way humans organize to use the available resources."[9] I remind them, quoting Liza Featherstone, that "blaming female breeders and our rugrats for societal ills is not new," but part of the centuries-long playbook of patriarchy.[10] Or I talk about the value, the meaning, the beauty of lives lived, even in terrible conditions. The people who lived (and died) at the centre of the Black Death, World Wars, famines and other horrific events: Would it have been better had those lives never existed? Or I quote Jedediah Britton-Purdy's essay on choosing to become a parent despite the threat of environmental destruction: "Part of the reason climate change is so terrible is the threat it poses to human life and culture, and I want to help them go on," not to make humans extinct.[11] Purdy avoids writing anything so glib as *the next generation is our best hope for establishing environmental justice* but I say those words to my students. Earth, our communities, all of us are damaged but that's nothing new. In Purdy's words: "A love for imperfect and impermanent things isn't a bad starting point for passionate democratic politics," which we'll need if things are to improve.

I talk about Naomi Klein's book for kids, which argues that youth-led mass collective action can be scaled up, intensified and, ultimately, transform the relationships between humans and the non-human environment. In *How to Change Everything: The Young Human's Guide to Protecting the Planet and Each Other*, Klein's vision of transformation centres on a Green New Deal that ends the burning of fossil fuel right now. It involves redesigning urban housing, extending Indigenous sovereignty over land, creating millions of environmentally safe good jobs. Klein is in favour of asserting democratic control over society's key productive resources (including land, power grids, communication networks and industrial factories) in ways that effectively overturn the foundations of capitalism. "It may seem overwhelming to some, because to do it effectively, we really do have to change everything."[12] Switching to cloth diapers and driving a Tesla isn't going to cut it. "But," Klein finishes the thought, "isn't that less overwhelming than the changes that will be forced on us if we fail to take well thought-out action against climate change?"

I'm moved by Klein's stories, and I accept the truth of her political analysis. Yet when I leave the theatrical environs of the lecture hall, moving from invigorating discussion with my students to sitting quietly in my office alone, when I drive home in the dark along the banks of the Grand River, my heart sinks. The reassurances I've given my students haven't stopped me from worrying.

Partly, that's because we are still so far from implementing and abiding by even the slimmest plan for ecological sustainability. But what I don't admit to my students – in fact, I hardly ever discuss it – is that the question of how I cope with parenting during the climate crisis triggers different parenting fears in me, fears that I experience as no less pressing a crisis. Because inasmuch as parenting has been about the rewards of offering unconditional love and watching human lives develop, my experience has been defined by fear – fear of something catastrophic happening to my kids now, or now, or now.

What if Winnie climbs the rail on the second-floor landing and drops fifteen feet to the floor? What if Gus drinks windshield wiper fluid? The parched riverbeds in my eco-nightmares are harmless; the real threats are the Grand and Nith Rivers as they exist today, flowing with fresh water. A friendly dare to dive in the river gone wrong; a drunken stumble next to the water at night; falling through thin ice. I grew up in a small town, and I survived each of these close calls. But not everyone does. My kids could be lost to the river (or the main road through town, or the railway bridge, or the cliffs over the Grand, or a house fire, or [thank god no longer]

SIDS, or meningitis, or a lightning strike). Yes, the eco-crisis might get them. But human life is fragile, vulnerable to far more pressing threats.

One week after Winnie was born in 2021, Jess detected a white spot in the centre of the baby's left pupil. In the weeks between the doctor referring us to a specialist and having that specialist offer her expert opinion, what does one do but google "white spot baby eyeball"? And so we learned of retinoblastoma, a type of eye cancer usually found in babies. Looking into the eyes of my small children has been among the most life-affirming things I've ever done. Jess has spent countless hours locking eyes with our children while feeding them at her breast. And in those first months, when we looked at Winnie, along with everything else we saw – all the love, the wonder, the universe in a tiny face – we also saw an unidentified white spot. Stop worrying, everyone said. The odds of it being something life-threatening are infinitesimal. Worrying doesn't change things anyway.

And it wasn't cancer. (It was a congenital cataract.) But, of course, we worried.

A month after moving to Paris, Gus woke in the middle of the night struggling to breathe. As I yelled to Jess to call an ambulance, my mind raced back to when I was twelve years old and learning CPR on the beach, and raced forward for a glimpse of the unthinkable. The ambulance arrived quickly. Gus was fine.

Tell parents these stories and they'll tell you similar and far scarier tales. Car crashes survived, concussions with lingering effects, death-defying fevers, dog attacks fended off. (When my brother was a toddler, he was mauled by a German shepherd.) My experience is the experience of every parent in my situation. The limits of what kids are capable of surviving are constantly being tested. Yes, kids are resilient, I'm not denying that. And most of the time kids pass the test. If they're not stronger as a result of surviving, at least they're alive at the end. But not every kid survives every test.

A report by Children First Canada says that accidents and preventable injuries – bike, car, poisoning, drowning, falling, fire – are the leading killer of Canadian children.[13] Comparatively high suicide rates reflect a "mental health 'crisis'" among young people across Canada.[14] There are tragedies. Last week, the daughter of a close family friend died from a drug overdose. And while fatal tragedies are exceptional, even more statistically unlikely among children in higher socio-economic groups, I never stop dreading them. When it comes to my experience of threats to my kids' lives, reason is no match for emotion.

At one time, I assumed that my worries over my children's lives (perhaps I mean their deaths) were uniquely intense and constant. But after talking to parents and reading parental memoirs, it's clear that I'm not alone in my obsessions. In a 1978 interview with the *Paris Review*, Joan Didion said: "The death of children worries me all the time. It's on my mind. Even I know that, and I usually don't know what's on my mind."[15] Didion's daughter was eleven at the time of the interview; she died at age thirty-nine, following nearly two years of illness, when Didion was seventy-one. In Keith Gessen's memoir-of-fatherhood *Raising Raffi*, he writes, "I do not know of a single parent who does not spend at least some time worrying about their child suddenly dying for no reason."[16]

In a *Paris Review* essay, Claudia Dey says mothers avoid talking about the central truth of parenthood: "with a child comes death."[17] Ideally, our children will live long, happy lives. What's certain is that one day they will die. And once you face up to the fact that mothers are "makers of death," writes Dey:

> Death slinks into your mind. It circles your growing body, and once your child has left it, death circles him too. It would be dangerous to turn your attentions away from your child – this is how the death presence makes you feel. The conversations I had with other new mothers stayed strictly within the bounds of the list [of practical

tasks]: blankets, diapers, creams. Every conversation I had was the wrong conversation. No other mother congratulated me and then said: I'm overcome by the blackest of thoughts. You? This is why mothers don't sleep, I thought to myself. This is why mothers don't look away from their children.

In *Harper's Magazine*, Perri Klass argues that anxiety and suffering over the death of children may actually have increased as child mortality has decreased. Children have always died and that has always been tragic. But prior to the middle of the twentieth century, and especially prior to 1900, it was common for families to lose a child. Not enough was known about sanitation, disease transmission and medicine was nothing like it is today. There are accounts of tremendous sadness in the loss of children by John Adams, Charles Dickens and others. There isn't, however, the suggestion that the event was unique or especially unexpected. "Modern parents who have lost children often describe feeling a sense of desolate isolation; there is no way for them to mention the dead child or the child's death in conversation without drawing attention to a tragedy so unthinkable that it crowds out all other subjects."[18] Death among young people in the West today is so rare that it's a truism to say, "Children are not supposed to die." Yet, Klass explains, you wouldn't have said that in the 1800s, when children died sometimes. Social and technological advances leading to a world where children are not supposed to die may be "our greatest achievement as a species." However, the rarity of children dying today may "make the grief harder to bear."

It probably also makes parents today more anxious. Endless safety features on offer, medical advice on-demand, even if they effectively guard against death, "can leave some parents terrified that somewhere along the way they will make a mistake, a wrong turn." We know anxiety is unhealthy. No one wants to be an anxious parent, and parents who are anxious are often made to feel as though it's a personal failing. Stop worrying! Just lighten up! Klass has compassion for the "impossible task" of averting "all possible threats and dangers," and "to do it without worrying." Every action encouraged to promote child safety is an invitation to imagine the death of the child.

It can feel as though parenting is a condition of chronic crisis, every moment the decisive one between your children living and dying. This doesn't mean that life is chronically bad. I'm not using the term "crisis" here to mean things already destroyed, sudden disaster or gradual decline. I'm using it to evoke the energy of the high-stakes moment. Crisis as the Hail Mary pass, the period of disruption in which danger and opportunity reside in equal measure, and from which redemption, even glory, are just as possible as failure and ruin. Whatever is human can be destroyed at any moment. And for parents who assume responsibility for the care of their kids, parenting may be the ultimate crisis, a crisis I doubt will ever end.

Why does my fear of the eco-crisis lead me to fear the immediate loss of my children? The fears are so different. The threat of ecological collapse is global, social and animated by models of probability. The threat of my children disappearing suddenly barely amounts to a speck of dust on the map. Because the possibility of my children's sudden death is highly unlikely and highly personal, it smacks of self-indulgence, not serious cause for concern. The one fear is rational; the second is irrational. Why are they fused in my mind?

Both are high-stakes fears – matters of life and death – with no guarantee that everything will be okay. They are mortal threats to what matters most to me without the possibility of vanquishing them. When and how will climate change end? How much time do we have to change course? What can we do, what can I do, that will make a lick of difference? It's impossible to say. That's terrifying.

While far less worrying statistically speaking, no less terrifying to me is that my children are fragile creatures in the here and now. As Montaigne (who lost five of his six children) once said: "Death can surprise us in so many ways."[19] No matter how dutifully and wisely I care for them, protect them, prepare them for the vicissitudes of a life, they are, because they're human, vulnerable to accident, to cruel acts of fate. In short, in both situations – global and local – while in very different degrees and with very different consequences, I lack a sense of control.

Psychologists studying the fear of uncertainty describe the pain of not knowing whether and when expectations will be violated. Crisis situations magnify the fear that worse is to come. By virtue of a situation being a crisis, it is not certain how things will turn out; the situation could go either way. The patient may recover or die. Earth may become largely uninhabitable during our children's lifetime, or eco-apocalypse may be averted. Our children might live long, happy lives, and they might be struck down tomorrow. We desire, maybe even fight for the crisis to be resolved this way rather than that way. But from within the chaos of the crisis, we cannot know how it will end.

In Corey Robin's book on the politics of fear, he notes that "fear is the first emotion experienced by a character in the Bible."[20] God creates and looks upon heaven and earth. Adam and Eve, before they're afraid, simply exist: they *are*. They see their garden. They name things. We're told: "They were not ashamed." Only once they face God's wrath, after eating the forbidden fruit, are they moved by experience. "Afraid, they are awash in experience, with God promising even more – for Eve the pain of childbirth, for Adam the duress of work, for both the dread knowledge of death."

Fear is among the great movers of Western literature. Fear of death (in *Hamlet*, in Keats, in DeLillo), of the supernatural (in *The Shining*, in *The Turn of the Screw*), of nature (in *Moby Dick*), of losing love (in *Romeo and Juliet*), of going mad (in Raskolnikov, in Poe's *The Tell-Tale Heart*), of failing to live well (in *Pride and Prejudice*, in Barnes's *The Sense of an Ending*), of history (in Rushdie's *Midnight's Children*, in Marías's *Thus Bad Begins*), of the future (in dystopias by Atwood and Butler).

The sociologist Frank Furedi argues that we live in a "culture of fear" today unlike at any point in history. People have always been afraid, of course – of death, of beasts, of the unknown. But, Furedi says, before the second half of the twentieth century, fear was something to be overcome, an emotion to be understood and managed within a broader cultural framework. The West's dominant *"cultural script"* from antiquity to the Second World War prized courage, heroism and respect for authority.[21] You learned to fear in relation to deep trust in human agency, creativity and progress. In the postwar decades, says Furedi, as the power of these cultural scripts has faded, fear has become the cardinal emotion, the language through which we express ourselves and know each other. "In the absence of a script offering a perspective on how to fear, *fear itself has become a perspective* through which life is interpreted."[22]

In our culture of fear, public space is choked by safety warnings – placards on the beach warning of "uneven surfaces," a sign at a cemetery warning that "all memorials have the potential to harm."[23] You get one group of parents worrying that the sun will give their kids skin cancer, a different group worrying that sunblock is poisoning their kids and a third group worrying that their kids aren't getting enough sun. People are compelled to act by "fear appeals" – constant warnings about "'risky behaviour,' 'unhealthy choices' or 'green sins.'"[24]

Furedi ends his book with a plea for "a less fearful future."[25] Only after the culture of fear is defeated will we be able to achieve our full human potential. If future generations are to escape the culture of fear, they must adopt (or return to) a new set of values: "particularly of courage, duty and judgement."[26] Dispense with trauma-clutching therapy. Stop labelling children as vulnerable or at risk. Celebrate the human capacity for problem-solving, resilience, bravery and social progress. Stop being so afraid of everything.

The authors of *Overcoming Parental Anxiety* use neuroscience to locate the problem in my lizard brain. We're hardwired to be on high alert, they say. You can't blame the amygdala for sending out "a cacophony of catastrophic cognitions," but you can train the prefrontal cortex to better manage your irrational fears, so "you will no longer be held captive by your worry thoughts but instead can choose how and when to respond to them."[27] My fear of uncertainty can be overcome through Cognitive Behavioural Therapy.

A study recently published in *Current Psychology* concludes that environmental activism can be "a buffer" against climate change anxiety. Loss of biodiversity got you down? Can't sleep because the ice shelves of Antarctica are splashing into the sea? Try this (actual examples of "actions" in the study): "Avoiding disposable, single use material [. . .] Educating oneself about environmental science [. . .] Spreading awareness (online, sending emails to elected officials, talking to other people about climate change."[28] Forget that the ecological impact of these actions is virtually nil, because "engaging in collective action may combat feelings of despair and helplessness and foster feelings of hope."[29]

From a neo-Marxist perspective, Richard Seymour writes about the importance of resisting "environmental melancholia" by remaining open to "enchantment" in a world disenchanted by Western philosophy and economics.[30] Not the enchantment of the pre-Enlightenment Old Regime that burned witches at the stake. And not the glib enchantment of David Attenborough's *Blue Planet*, which portrays nature as enchanting by narrating the lives of whales and monkeys as though animals were driven by the same desires and fears that drive humans. Seymour's enchantment comes down to believing in the possibility of the transformation of the world at a scale and of a depth that we cannot possibly fathom from where we stand. While living in the world, we must cultivate "unworldliness." He recognizes this is a type of faith. It involves making "a secret wager that other worlds can exist." More doctrinaire radicals than Seymour assure me that "the people united can never be defeated."

In *Society of Fear*, the sociologist Heinz Bude recommends the anxious among us learn from Mikhail Bakhtin's work on ribald laughter unleashed during Medieval carnivals. "Medieval man perceived laughter to be a victory over fear."[31] Fear of authorities, hunger, natural disasters and death was laughed away in festivals celebrating bodily pleasure, blasphemy, inverted social relations, the grotesque.

Thomas Doherty, an Oregon-based psychologist specializing in environmental anxiety, says that beyond making green consumer decisions, parents can do little about climate change other than "simply to bear witness to the issue."[32] He cautions parents against assuming it's their responsibility to fix the problem. When it comes to facing the eco-crisis, says Doherty, the key thing for parents is to act knowing "that even with the dire scenarios predicted, there are going to be good days in the future. There will be bad days: disasters, fires, floods. But that doesn't mean that there won't be sunny, good days for ourselves and for our families."

It's difficult for me to adopt any of these approaches for overcoming my greatest fears. Not so much because I think they wouldn't work but because each starts from the premise that my fears are overblown, irrational, something to be gotten over. The fact that my fears are comprised of a jumble of facts and gut feelings, statistical probabilities and what Lee Clarke calls "possibilistic thinking,"[33] real threats and worst-case scenarios, doesn't make them less valid, less meaningful. They are meaning itself. I question the honesty, at least the humaneness, of others who don't feel as I do. If clinging to my fears is absurd, don't blame the dysfunction of an individual mind. Blame the absurdity of the situation we're in.

Albert Camus, the Algerian-French writer and Resistance fighter, described the absurd as "the confrontation of the irrational and the wild longing for clarity whose call echoes in the human heart."[34] Humans want full understanding and control; but the world is not fully knowable, and it's ruled in large measure by chance. Camus called existence absurd because of the contradiction between our big questions – What is the meaning of life? How do I know? How should I live? – and the impossibility of answering those questions by reference to any external system, religious or otherwise.

Whatever else it is, it is absurd that we know that burning fossil fuel destroys the basis of life, even as we burn it in world-destroying batches every minute. It is absurd to be hurtling toward eco-apocalypse and continue doing the very things driving us there. Politicians set eco-targets – the Kyoto Protocols, the Paris Agreement – and smash through their limits without penalty, evidently without shame and without any action to suggest next time will be different. Heads of state, corporate bosses, philosophers and religious leaders tell us the environment is sacred, life matters. Meanwhile, forests are felled, the oceans are poisoned and a million and one luxury resorts go on washing unused towels. A World Cup of soccer is held in the desert. It's absurd.

As a parent, it is absurd that the heart of my existence, the meaning and material of my life, my kids, could be gone in a flash, yet I must act as though this isn't the case, and I'm encouraged to stop thinking about it. It's also absurd – cosmically unfair, cruel by chance – that I have the means, time and space to experience such joy while billions of humans never will. It's absurd that I worry so much about my family when the risks to my little kin group are so small. That I can know all these problems and envision how they might be fixed while feeling incapable to do anything about it, this is the experience of the absurd.

The Marxist in me says I'm mystifying matters. I don't need Camus's fancy language to name my fears or to describe what it means to be human. The irrationality and suffering, both social and personal, that I've called absurd are neither eternal nor random. They are manifestations of the class struggle playing out at different levels of experience. The world is burning in spite of what we know about climate change because it's economically rational for corporations to burn it. And it isn't "us," humans collectively, doing the burning, but the executives of private interests.

If revolutionary class struggle ends global inequality, I will be as ecstatically happy as the next comrade. I have no trouble accepting that Camus's condition of absurdity is a symptom of modernity, not applicable to the human condition in the abstract across millennia. But at this moment in the climate crisis, amid my crisis of parenting, I talk to my kids about the importance of environmental stewardship while the Government of Canada buys an oil pipeline. The Ontario government rejected referendum results it promised to honour in order to greenlight building more subdivisions that only a fraction of the population could buy into on some of the province's most biodiverse land. In 2021, the ocean caught fire after gas leaked from an underwater pipeline in the Gulf of Mexico. Thanks to increasingly invasive extraction technologies, the United States has become the world's leading oil producer while it claims also to lead the green revolution.

Philosophers and activists have dismissed the absurdist orientation for provoking despair. Camus would say they've missed the point: "To observe that life is absurd cannot be an end, but only a beginning. . . . What interests me is not this discovery [of life's absurd character], but the consequences and rules of action we must draw from it."[35] In Camus's own biography, embracing absurdity involved exposing abuses of power, supporting free speech and fighting oppression. The great philosopher of the absurd was also the chief editor of *Combat*, a clandestine newspaper of the French Resistance. His absurdist orientation involved savouring sensual pleasures – the sun on our skin, fresh air (and French cigarettes) in our lungs – because the time for living out our humanity is now, not in a future perfected society, not in the afterlife (which doesn't exist). In an article on artists of the absurd, Pedro Querido ascribes to Camus "the disruptive fervor, and the inescapable imperative to resist, to subvert, and never to surrender, be it to soul-eroding hyperrationality or to the absurdity of life (generally speaking or in a particular socio-political context)."[36]

What does an absurdist disposition do for me as I grapple with my fears of uncertainty, fears that manifest both in the context of parenting amid eco-crisis and in the ultimate crisis of parenting more broadly? It creates space to attend to the contradictions of crisis. It means being honest about my fear, fatigue, rage and periodic despair without embracing defeatism. Put the other way around, it means acting with purpose, with the intent to effect change, both privately and politically, without illusions about the consequences of my actions. Being paralyzed in fear or despondency in the face of death is not truly living. Yet it's also delusional, inauthentic, untrue to live as though death is of no concern, can be put off for later consideration, forgotten about. Montaigne again: "We do not know where death awaits us: so let us wait for it everywhere."[37]

An absurd orientation maintains the tension between our love of life and the guarantee that we will die. We recognize that human knowledge and power is limited, but we strive ceaselessly to learn and to transform the world. We are committed to fairness, justice and progress, while seeing brutality everywhere and knowing well the cruelty of chance. We do this not in spite of our humanity or to somehow escape humanity's flaws. We do it in order to live as our most authentic selves. The absurdist stance means letting go of every false hope, every empty platitude of reassurance, while proceeding to act with the intent of resolving the crisis this way instead of that way.

I'm not the first to hold up an absurdist lens to the climate crisis. In the *Journal of Climate Change and Health*, Jacob Fox describes the "irrational dissonance between our demands for climate action and the inertia, cynical or otherwise, of those with the power to enact them. Moreover, the climate emergency, like the specter of atomic war that haunted Camus's generation, may hasten individual death and the demise of our species. This renders more urgent both that [human] desire for meaning and the temptation to resign to our mortality."[38] The ethicist Richard Gibson grants that living through the climate crisis is absurd, "but so is everything else we do. Indeed, the universe itself, with its gravity, black holes, hummingbirds, earthquakes, x-men, and suntan lotion, is nothing but absurdity. So, why should we think that our lives should be any different? Why should the absurd task of saving the world from climate change be notably outrageous compared to everything else?"[39]

Blanche Verlie's book *Learning to Live with Climate Change* talks of the "guaranteed uncertainty" in our relationship with the environment.[40] Living with climate change "is not about becoming resigned to climate change, giving up or thinking that it is too late to do anything."[41] It does, however, mean recognizing that "future ways of living-with will be radically different to those we have come to know and/or love, and involves grieving for the losses we are already experiencing and those that are yet to come." It means "attuning to the uncanny, disorienting and devastating experiences of encountering climate anxiety."[42] In their own ways, both Camus and Verlie counsel us to nourish the forces of life-making by keeping our destructiveness and mortality at the forefront of our mind, even as we strive to live as long and richly as possible.

Part of being human, core to the experience of the absurd, is the impulse to resist. Resist climate change, fascism, corruption and lies, resist threats to our loved ones, resist death itself. But we resist with no guarantee. Worse: we resist with the guarantee that we and everyone we love will die, and it will be as though none of us were ever born. In 2025, we resist when the forces of ecological death are still far stronger than the forces of life-making. Historians talk about Europe in 1914 sleepwalking into war. Will we wake up before the sixth extinction is irreversible? I don't know. None of us does. But I'd rather embrace not-knowing as part of our clear-eyed struggle with absurdity than dismiss it as a catastrophizing remnant of my lizard brain or try calming myself with political platitudes about the forces of justice being undefeatable. History knows that the good guys don't always win.

There's no quelling my fear that harm will come to my children. I doubt I'll stop worrying about their fates, nor will I give up my effort to defend against every imaginable risk, but I understand that the pumping of their hearts is beyond my control. Absurd reasoning involves struggling for humane resolutions to the crisis of parenting, the crisis of parenting during the eco-crisis, while, like Camus's Sisyphus, knowing "the whole extent of [our] wretched condition."[43]

One of my favourite things to do with my kids is throw stones in the rivers near our home. We walk to the boat launch in the centre of town and face the Grand River from large slabs of limestone. I ask Winnie which sound she prefers: the *bloop* of a single small stone dropping into water at our feet or the *whish* of dozens of tiny rocks splashing at once after my throwing them skyward. From a clearing in Barker's Bush, Gus fails to skip stones across the surface of the Nith River, then searches the shoreline for crayfish.

When I'm at the rivers with Gus and Winnie, I think about our relationship to the earth and the strangeness of the passage of time.

I sense the truth in Heraclitus's dictum: You cannot step in the same river twice. The water is moving and we are growing older. Flow, change, mutability, ephemerality. This is the river; this is us.

Of course, as Heraclitus well knew, you *can* step in the same river twice. The river moves, but it stays still. And when I look at the Grand, I picture people stepping in the same spots for centuries. Haudenosaunee hunters, European explorers, labourers, tourists, fishers, kids, my kids: Gus and Winnie. History is here, along with the fish and bugs and green goose shit. And the Grand isn't just here in Paris, it's downriver in Brantford, too, then in Ohsweken on the Six Nations of the Grand River reserve, through Caledonia, Cayuga, Dunnville and Port Maitland before flowing into Lake Erie. Imagine where the river starts. Bubbling up from the ground? Liquid thread connecting Belwood to Elora to Cambridge to Glen Morris and a million nameless places along the way – the rocks that herons stand on, the beer bottles smashed against bridge abutments, crayfish graves, a sunken shopping cart, campsites and osprey nests.

No doubt the chemical composition of the water is vastly different today than it was before the town of Paris, Ontario, was established and not only because most of the gypsum has been removed. Paris was home to large textile mills from 1874 until late in the twentieth century. Exploited for profit, protected out of love, the nature of Paris's rivers is more like the manicured nature of Niagara Falls and Central Park than the comparatively untouched wilderness of the tundra, the mountain, the bottom of the ocean. But it's nature, nonetheless. Paris's rivers form the core of the regional ecosystem. And while I tend to think of the rivers' life-giving power when I'm standing next to them with Gus and Winnie, rivers symbolize death, too. The River Styx. Michael, row your boat ashore. In the New Hamburg area of the Nith alone, twenty-seven people have drowned since 1871. A fact and a symbol of life, death, time passing, our animalistic being: no wonder my kids scuffing through the parched riverbed is what I see when I think about the environmental crisis.

NOTES

1   David Wallace-Wells, *The Uninhabitable Earth* (Tim Duggan Books, 2020), 3.

2   David Wallace-Wells, "The Uninhabitable Earth," *New York*, July 10, 2017, https://nymag.com/intelligencer/2017/07/climate-change -earth-too-hot-for-humans.html.

3   Quoted in Sean Illing, "It Is Absolutely Time to Panic About Climate Change," *Vox*, February 24, 2019, https://www.vox.com/energy -and-environment/2019/2/22/18188562/climate-change-david -wallace-wells-the-uninhabitable-earth.

4   Quoted in Heather Avery, "Old Crow, Yukon, Declares Climate Change State of Emergency," *CBC News*, May 21, 2019, https:// www.cbc.ca/news/canada/north/old-crow-climate-change-emergency -1.5144010.

5   Britt Wray, introduction to *Generation Dread: Finding Purpose in an Age of Climate Crisis* (Knopf, 2022), Kobo.

6   Wray, "The Psychoterratic State," chap. 1 in *Generation Dread*.

7   Wray, introduction to *Generation Dread*.

8   Wray, *Generation Dread*.

9   Meehan Crist, "Is It OK to Have a Child?," *London Review of Books*, March 5, 2020, https://www.lrb.co.uk/the-paper/v42/n05/meehan -crist/is-it-ok-to-have-a-child.

10   Liza Featherstone, "Don't Blame the Babies," *Jacobin*, March 15, 2019, https://jacobin.com/2019/04/children-climate-change-family-guilt.

11   Jedediah Britton-Purdy, "The Concession to Climate Change I Will Not Make," *Atlantic*, January 6, 2020, https://www.theatlantic. com/science/archive/2020/01/becoming-parent-age-climate-crisis /604372/.

12   Naomi Klein with Rebecca Stefoff, "Changing the Future," chap. 7 in *How to Change Everything: The Young Human's Guide to Protecting the Planet and Each Other* (Puffin, 2021), Kobo.

13   *Raising Canada 2022: Top 10 Threats to Childhood in Canada* (Children First Canada, 2022), https://childrenfirstcanada.org/wp-content /uploads/2022/09/RC2022_CFC_RC-Report_09-02.pdf.

14 Robert Cribb, "More Than 5,800 Young Suicides Across Canada Signals Mental Health 'Crisis,'" *Toronto Star*, September 16, 2018, https://www.thestar.com/news/investigations/2018/09/14/youth-suicide-is-like-a-cancer-experts-warn.html.

15 Joan Didion, interview by Linda Kuehl, "The Art of Fiction No. 71," *Paris Review*, no. 74 (Fall–Winter 1978), https://www.theparisreview.org/interviews/3439/the-art-of-fiction-no-71-joan-didion.

16 Keith Gessen, "Zero to Two (The Age of Advice)," in *Raising Raffi: The First Five Years* (Viking, 2022), Kobo.

17 Claudia Dey, "Mothers as Makers of Death," *Paris Review*, August 14, 2018, https://www.theparisreview.org/blog/2018/08/14/mothers-as-makers-of-death/.

18 Perri Klass, "When Children Die," *Harper's Magazine*, June 2021, https://harpers.org/archive/2021/06/when-children-die/.

19 Michel de Montaigne, "To Philosophize Is to Learn How to Die," in *The Complete Essays*, trans. M.A. Screech (1588; Penguin, 2003), 94.

20 Corey Robin, *Fear: The History of a Political Idea* (Oxford University Press, 2004), 1.

21 Frank Furedi, *How Fear Works: Culture of Fear in the Twenty-First Century* (Bloomsbury, 2018), 15, italics in original.

22 Furedi, *How Fear Works*, 141, italics in original.

23 Furedi, *How Fear Works*, 208.

24 Furedi, *How Fear Works*, 128.

25 Furedi, *How Fear Works*, 237.

26 Furedi, *How Fear Works*, 252.

27 Debra Kissen, Micah Ioffe and Hannah Romain, introduction to *Overcoming Parental Anxiety: Rewire Your Brain to Worry Less & Enjoy Parenting More* (New Harbinger, 2022), Kobo.

28 Sarah E.O. Schwartz et al., "Climate Change Anxiety and Mental Health: Environmental Activism as Buffer," *Current Psychology* 42 (February 2022): 16716, https://doi.org/10.1007/s12144-022-02735-6.

29 Schwartz et al., "Climate Change Anxiety": 16718.

30 Richard Seymour, "Unworldliness," in *The Disenchanted Earth: Reflections on Ecosocialism and Barbarism* (Indigo Press, 2022), Kobo.

31 Heinz Bude, *Society of Fear*, trans. Jessica Spengler (Polity, 2018), 114.

32 Quoted in Katie C. Reilly, "A Calm Guide to Climate Anxiety for Parents," *Parents*, March 30, 2022, https://www.parents.com/parenting/better-parenting/green/a-calm-guide-to-climate-anxiety-for-parents/.

33 Lee Clarke, "Possibilistic Thinking: A New Conceptual Tool for Thinking About Extreme Events," *Social Research* 75, no. 3 (2008): 676.

34 Albert Camus, "The Myth of Sisyphus," in *The Myth of Sisyphus and Other Essays*, trans. Justin O'Brien (1955; Vintage, 1991), 21.

35 Quoted in Robert Zaretsky, *A Life Worth Living: Albert Camus and the Quest for Meaning* (Harvard University Press, 2016), 17.

36 Pedro Querido, "From Kharms to Camus: Towards a Definition of the Absurd as Resistance," *Modern Languages Review* 112, no. 4 (2017): 792.

37 Montaigne, "To Philosophize Is to Learn How to Die," 96.

38 Jacob Fox, "Camus and the Climate Crisis," *Journal of Climate Change and Health* 6 (May 2022): 2.

39 Richard Gibson, "COP26: What's the Point?," *Prindle Post*, November 5, 2021, https://prindleinstitute.org/2021/11/cop26-whats-the-point/.

40 Blanche Verlie, *Learning to Live with Climate Change: From Anxiety to Transformation* (Routledge, 2022), 51.

41 Verlie, *Learning to Live*, 113.

42 Verlie, *Learning to Live*, 59.

43 Camus, *Sisyphus*, 121.

BLACKOUT

What I want you to know is that we're okay. Hurting
but okay. We're surviving, though it's true,
we don't know what that means, exactly.
– Franny Choi, "Dispatches from a Future Great-Great-Grand-
daughter"

In late summer 2002, not long after my twenty-third birthday, I
moved from Toronto to Saskatoon to start a master's degree. On
the prairie I lived alone, on the top floor of a three-story apartment
building. My small balcony faced a black iron bridge spanning the
South Saskatchewan River. During my twelve months in Saskatoon, I blacked out from drinking several times a week.

I'd wake with no idea how the night before ended, who I'd
been with or how I'd got home. Or I'd wake in a house I swear I'd
never been in before. I'd recall speeding along a dark highway, but
not who was driving the car. I'd guess at my movements from the
previous night based on physical evidence found in the morning.
Whose backpack is this? Why is my oven filled with bowls and
plates? I must have left the apartment because my coat isn't here.
Like all blackout drinkers, I had an active life that was completely
unknown to me. I perfected what Leslie Jamison calls "the post-
blackout processing session, letting someone tell me what I'd done
and then helping him figure out why I might have done it."[1]

In an alcoholic blackout (which is different from passing out),
drunks go on acting in the world, but later have no recollection of
doing so. Alcohol levels temporarily shut down the brain's ability
to record long-term memories. Drinkers won't remember what
happened during a blackout, not because they aren't trying hard

enough, but because there's no memory to recall. Memories were never made.

The godfather of blackout research, psychiatrist Donald W. Goodwin, once got a group of men in a laboratory to drink twenty ounces of bourbon in under five hours. He asked his drunk research subjects if they were hungry. He pointed to a covered saucepan. Lifting the pan's lid, Goodwin revealed three dead mice. The intoxicated men had forgotten the episode thirty minutes later. The next day they recalled nothing. Goodwin declared success: "We had produced experimental blackouts."[2]

Goodwin, Crane and Guze's classic study of one hundred hospitalized alcoholics distinguishes between "fragmentary" and "*en bloc*" blackouts. In a fragmentary blackout, partial memories can later be recalled, though not clearly or continuously. One research subject said recalling a partial blackout was like "remembering a dream."[3] By contrast, writes Goodwin's team, after an en bloc blackout, "no amount of memory jogging, either by the person in an effort to remember or by friends telling him of the event, dispelled the amnesia." I'm surprised that only sixty-four of the one hundred alcoholics reported experiencing blackouts. Recent studies suggest that more than half of US college students have blacked out from drinking at least once. A study of 954 students with "problem alcohol use" in Canada and the US reports that around 7 percent of those surveyed experienced six or more blackouts in the previous year.[4] My conservative estimate is that, between my sixteenth birthday in 1995 and my forty-fifth birthday in 2024, I blacked out, fragmentarily and en bloc, more than one thousand times.

One morning during my winter in Saskatoon, I woke with my shirt on properly and my pants around my ankles. This was odd. As a rule, after a blackout, I woke either fully clothed or totally naked, with my clothes folded neatly next to the bed. More strangeness outside the bedroom: the main door to the apartment was ajar. I'd never left the door ajar overnight.

I walked twenty minutes in the crunchy snow to my coffee shop job. I opened the café and turned on loud music. I set up chairs, sliced vegetables and drank cup after cup of black coffee. My hands shook, first from the hangover, then from too much caffeine.

A coworker joined me an hour later. We commiserated over the cruelty of hangovers. I told her I'd been to the Yard and Flagon and must've closed the place down. "Must have?" she asked. We laughed. "You know how it is," I said. We laughed.

At lunch, a woman arrived who came in most days around that time. She had a fake tan and wore a blue parka. She and I often flirted, though she was twice my age, married and had three kids. I gave her my sunniest smile despite being close to vomiting. She leaned against the counter and asked, "So, how are *you* feeling today?"

I took a deep breath, wiped the counter with a cloth and said, "Not the best I've ever felt, to be honest."

She snorted. "Yeah. I'll bet."

"Oh shit." I looked up. "Did you see me last night?" I scanned my memory for an image of her at the pub. Blackness.

The expression on her face changed suddenly, from teasing to confusion to what looked like hurt.

My face turned cold.

"I was really fucked up last night, Caroline," I said. I had the sensation of falling through a hole in the centre of my chest. "We started drinking Jim Beam at like 2:00 p.m. That stuff erases my brain."

"You seriously don't remember?" Her incredulity frightened me. I was starting to panic. The ambient noise of the café disappeared.

"I'm so sorry, Caroline. What did I do?" Now I pictured her in a blue minivan parked in front of the pub. I pictured her approaching me in the pub, saw my unlaced boots on the floor of her passenger seat as she turned the minivan left into my parking lot. But I was making this up. Right? Surely I would remember if she drove me home less than twelve hours ago. I searched for more images. Blackness. But images kept appearing, whether memories or fabrications. Had she been in my apartment? I remembered the unlocked door, my pants around my ankles in the morning. I tried to sound relaxed, but my voice was strained: "Wait, did you drive me home? I'm so sorry, Caroline, I honestly don't remember anything."

She smirked, but not happily. "You honestly don't remember." Now her tone was declarative, though still she sounded dubious.

I gave up trying to seem anything other than as desperate as I felt. I told her I'd been at the pub for hours. I'd been drunk for days. Everything was blending together. I assumed I'd been at the pub until closing, then walked home, as I'd so often done, but really, had no idea. I told her I was starting to get scared.

This seemed to please her, but not in a way that included me in the fun. I told her about the door left open in my apartment. Said I was worried about who'd been inside. She said, "Look, you don't remember? That's your problem."

She carried her food to a table against the wall. Between serving customers, I sat across from her and begged her to tell me what she knew about the previous night. "Caroline, this is serious for me. I'm so sorry to be like this. Please. I'm going to pull myself together. Just tell me what happened, and I'll never get that drunk again."

"Yes, you will," she responded. "I know people who drink like you do. Makes no difference what you say. You'll do this again, and again, and again. You'll do this your whole life." I went to the bathroom and cried.

In Sarah Hepola's *Blackout: Remembering the Things I Drank to Forget*, she writes that even a sober person can't always tell when their drunk friend has blacked out. Watch me, Hepola says, in the middle of a blackout. You'll see "a woman on her way to somewhere else, with no idea her memory just snapped in half."[5]

The imagery resonates with me because it evokes a before and after, a decisive break in being. After blacking out once at the Imperial Pub in Toronto, I returned to consciousness a few blocks away, lying on damp pavement in an alley. My shoes and eyeglasses were missing. I don't know where I'd been. (I never found the stuff I'd lost.) When I fell into the blackout, I went elsewhere. I'd return changed.

Hepola describes coming out of a blackout while on a magazine assignment in France. The last thing she recalls is exiting a cab after a night of heavy drinking with a friend. She walks with purpose from the taxi to the stairs in her hotel lobby, then the curtain falls: blackness. "When the curtain lifts again, this is what I see. There is a bed, and I'm on it. The lights are low. Sheets are wrapped around my ankles, soft and cool against my skin. I'm on top of a guy I've never seen before, and we're having sex." She seems to be enjoying it. The man seems to know who she is. She's not scared, just clueless. They finish and she lies in the crook of his arm thinking, "*How did I get here?*"[6]

How does anyone get here? Why drink to the point of blackout? Hepola says she drank to ease social inhibitions, obliterate her body and forget the harm her drinking caused. I, too, drank to manage sexual anxieties. I also relished dancing on a chair at the party's centre, and I knew drinking would get me there. Yet my reasons for drinking to oblivion never seemed rational to me. I don't know how to describe them without sounding comically abstract, arrogant, absurd. I drank to . . . what? . . . I don't know . . . to blackout in order to feel nothing. Or to feel everything without distinction. I wanted to experience total control over all of God's green earth. To sense meaningful changes of space, time and self. To approach the thrill of catastrophe, death. In Marx's *Economic and Philosophic*

Manuscripts of 1844, he explains that capitalism alienates people from our capacity to labour – the core of what Marx calls our "species-being" – as well as from one another. Capitalist wage-labour "estranges man's own body from him, as it does external nature and his spiritual essence, his *human* being."[7] Amid the life-destroying demands and violence of capitalism, who doesn't ache to feel whole, alive, free, of significance, of value? Who isn't, at some level, trying to escape?

Why did I drink that way? is a good question; one I've spent many hours in therapy trying to answer. But six years into my latest stretch of sobriety, it's not the question that troubles me most. When I lie awake at 3:00 a.m., I'm thinking: What do I do with all those blackouts? How do I relate to so much lost time? Time lost both in the sense that the blackouts are in the past and therefore unchangeable and the sense that when I lived them, they were, in that moment, already gone. How do I live with that night with Caroline (not with Caroline?), and so many others like it?

I've tried striking these dark episodes from the autobiography in my head on the grounds that I can't claim something I never knew. (Brain doctors are split on whether blackouts are legitimate legal defences.) But when I ignore the blackouts of my past, I'm unable to conceive of myself as someone whose views and actions matter today. The philosopher Martin Hägglund might say that my blackouts have broken "the fragile coherence of who I am trying to be," as a teacher of social justice, a socialist whose comrades can trust him, a parent whose guidance is more than hypocrisy.[8] It's necessary to face up to my history of blackout drinking if I am to find what Hägglund, in *This Life,* calls "spiritual freedom." Achieving biographical coherence (or at least, comprehensibility), which is an aspect of spiritual freedom, "requires that the agent in question can ask herself how she should spend her time and be responsive to the risk that she is wasting her life." What of the wastelands, black holes, in my past?

What about whole periods of drinking? Not blackouts technically speaking, but years, decades dominated by addiction, which I've also come to think of as blackouts of a kind. Huge swaths of time lost to my obsession with, consumption of and sickness from alcohol. When the poet Patrick Lane first returned from rehab, he said he "stepped back into the world after an absence of forty-five years of addiction."[9] The blackout is the condensed form of my twenty-five-year crisis of alcoholism, but what I've lost to drinking far exceeds all those memories never made. Staying sober, living well, avoiding the worst of this familiar crisis, requires a more considered approach to the darkness in my past. What might that involve?

Steps eight and nine of Alcoholics Anonymous involve listing "all the persons we had harmed" while drinking and making "direct amends to such people wherever possible."[10] But the blackout, by virtue of being the blackout, means that its harms are often untraceable. I once came out of a blackout early in the morning while entering a house that wasn't mine. I remember a woman with white hair saying, "No, no, you can't come in!" I'd somehow made it through her front door. She had an accent. Portuguese? While she waved a rolled-up magazine at me, I turned and walked back outside. Next thing I remember, the sun was up, and I was walking past a school near my house. How do I make amends to that white-haired woman when I don't even know what street she lives on? Encountering the "blank spaces of the historical archive" leads Saidiya Hartman to ask: "How does one write a story about an encounter with nothing?"[11] My ethical-methodological dilemma: Is it possible to rescue anything from the dead zone of blackout?

Until recently, I would've said no. I cultivated my guilt, my shame, knowing that while these feelings don't fix anything, I deserve their punishment. I'd do my best to forget the wreckage I'd left behind and try not to drink today. I made amends to some people for some wrongs I recalled. I told myself: the harm is done; that's tragic. Suffer, and go on.

A few years ago, while doing research for this book, I rediscovered imagery and language for redeeming the past that have me rethinking how to live with my blackouts. The source was unexpected. Not AA or the field of psychology's personal crisis literature. It's the German writer Walter Benjamin's 1940 essay "Theses on the Philosophy of History." Benjamin wrote it shortly before committing suicide while fleeing the Nazis at the Franco-Spanish border.

I'm not the first to apply Benjamin's ideas to alcoholism. (Considering his exalted status among academics, artists and activists, it's hard to imagine being the first to do anything with Benjamin's ideas eighty years after his death.) A 2013 article in *Contemporary Social Science* argues that "the rooms of recovery from alcoholism" are among the last remaining in which to find storytelling in the Benjaminian sense.[12] Like storytelling, alcoholism exists "in relation to dying."[13] And recovery, at least in Alcoholics Anonymous, is a process not of information-exchange, but of storytelling: the collective passing down of, and absorbing, embodied experience. My use of Benjamin supports the thesis of that article. But I'm interested in different aspects of his political thought: specifically, Benjamin's argument that revolutions have the power to reopen history. The theory of crisis and redemption that Benjamin develops in his final essay leads me to wonder whether time lost to alcoholic blackouts can be redeemed in the moment of quitting.

I'm aware that Benjamin's ideas about history and my personal biography operate on vastly different scales. My integration of them may, at first, seem mismatched, if not self-indulgent. Considering Benjamin's own experiments with collage, pastiche, dream interpretation, vaulting between levels of abstraction, drawing links between personal and political, it's possible that he would've savoured my audacious (grandiose?) approach. Certainly, he'd agree with my reflections below about the importance of becoming heirs to the revolutionary spirit in all aspects of our lives. Benjamin is the one who taught me "to brush history against the grain."[14]

Benjamin wrote his theses on history a few months before Hitler took Paris. He handed the essay to his friend Hannah Arendt, the eminent philosopher and fellow German Jew, also exiled in the French capital. With Nazis approaching Paris in June, Benjamin fled south. Stripped of his German citizenship, refused status elsewhere in Europe, he was poor and stateless.

In September, he crossed the mountains into Spain with a group of Jewish refugees seeking passage to America. Benjamin was denied entry to Spain because he didn't have a French exit visa. Under guard in the small town of Portbou, he learned that General Franco's police would deport him back to France. Benjamin's final note begins: "With no way out, I have no choice but to end it."[15] On September 25 or 26, he swallowed morphine pills and died. Whatever else the theses are, then, they are Walter Benjamin's final dispatch from what fellow revolutionary writer Victor Serge called "midnight in the century."[16]

The fourteen-page essay consists of eighteen numbered theses and two addenda. Its critical thrust is directed toward conventional thinking about what history is and the process through which social change happens. Specifically, Benjamin condemns liberal, Communist and social democratic allies for failing to comprehend the rise of fascism. He blames the failure on the widespread false assumption that history is the story of progress. Liberals who viewed technological development and the spread of markets as democratizing. Stalinists who saw dealmaking with Hitler as a step toward Soviet dominance. Social Democrats in Germany's interwar years who accommodated Nazism on the belief that the working class "was moving with the current" of socialism.[17] All of them were delusional. Because they believed in the steady progress of history, they saw historical advances where, in fact, fascism was cohering into a force of world domination.

In Thesis IX, Benjamin envisions the angel of history flying into the future with a face turned toward the past: "Where we perceive a chain of events, he sees one single catastrophe which keeps piling wreckage upon wreckage and hurls it in front of his feet."[18] The angel wants to help, to stop the carnage, to change the future, but his wings are pinned back by the force of the storm propelling him forward. "This storm is what we call progress." In fact, Benjamin famously says in Thesis VII, "there is no document of civilization which is not at the same time a document of barbarism."[19] Europe's greatness was built on working-class suffering at home and the plundering of the Global South. American democracy was dependent on genocide and slavery. Scientific breakthroughs are turned into weapons of mass death.

The idea that history is progress, each moment and event added to the last, evolving toward a higher goal, requires the fantasy of what Benjamin calls "homogenous, empty time."[20] In empty time, history is smooth, continuous, singular. Change is superficial. Hierarchies of power are persistent. Innovations are made – the nation-state, the steam train, the machine gun, the internet – but everything happens "in an arena where the ruling class gives the commands."[21] Evoking Nietzsche's concept of "the eternal recurrence of the same," Benjamin says that in empty time, no matter variations in official histories, life unfolds through the eternal return of ruling-class power and subordinate-class oppression.

Benjamin's notion of empty time helps convey my experience of addiction. In periods of active drinking, what I think of as *life inside the drinking frame*, I see the clock turn but change amounts to superficial variations of existence. Perhaps from the outside my life appears as progress. I fall in and out of love, read and write, move apartments. I get older, more credentialed, have children. But life inside the drinking frame is one single catastrophe: sick from consuming and from not consuming. Sneaking drinks on the subway, at school, at work; puking up hangovers all over the city.

Waking with no memory of the previous day. Gulping vodka from a water bottle hidden behind the bathtub while my girlfriend waits upstairs at the Scrabble board.

It's the eternal recurrence of the same: blackout. No matter how the day begins, what city I'm in, what's the occasion, I'm blacked out before the day ends. My sixteenth birthday: blacked out in my brother's car on the way to the party. My first conference as a PhD student: blacked out in a jail cell at the police station. My son's baby shower: blacked out opening gifts on the couch. The days were special in name only. In reality, each was empty time sick with alcohol, no different than every other day drunk alone in my kitchen. Inside the drinking frame is the appearance of development where, in fact, time is flattened by blackout. Ask Leslie Jamison, a recovering alcoholic who wrote a book about drinking memoirs: *Addiction is just the same fucking thing over and over again.*[22]

Escaping from empty time requires genuine change. This type of change, explains Benjamin, happens through radical interruptions, discontinuities, ruptures in the status quo. In his words, making history is "to make the continuum of history explode."[23] Exploding the continuum of history is the work of revolution. Overthrowing elites, redistributing land, democratizing power across society. In revolt, the oppressed shatter relationships of control, as well as the confines of empty time.

Time is so radically altered in revolutionary periods that Benjamin likens revolution to the coming of the Messiah. There was before; and there is now. The crisis deepens, then detonates. The world is new. The Messiah's arrival means not only the fall of the reigning earthly powers, but the beginning of "time filled by the presence of the now."[24] Now-time is the time of true freedom, the fulfillment of human potential. It's characterized by the conscious, self-directed activity of humans working collectively to meet the needs of all.

Reading Benjamin on revolution was the first time I identified with a description of what it feels like to quit drinking. Since the first time I quit, in 2005, I've quit eight or ten times. And while I've never strung together seven sober years, my episodes of abstinence have been long enough to teach me that recovery takes quiet, agonizing, unending work. The moment of quitting itself, however, that's always felt to me like an explosion. In the moment of quitting, the revolution kicks off. Time is ruptured. The Messiah enters the gate.

The last time I quit, the blast occurred in my living room. My son was three months old. My dad was visiting for the weekend. On a day I'd promised not to drink, my partner found me secretly pouring a can of real beer into my empty can of non-alcoholic beer. I was already buzzed from chugging vodka that I'd hidden in the basement ceiling. My partner, understandably, was furious. But she'd been furious before. We'd had a bigger fight about my drinking only the previous month. It ended with my promising to abide by new rules but also with me no less convinced of my right and need to drink, and the fact that I would continue to do so. Yet now, weeks later, on a sunny May day when I'd risen no less attached to blackout, I'm suddenly overwhelmed by the realization: I need to quit. And I will quit; am quitting.

I sit on the couch and tell my partner and my dad. The walls in the living room move away from me. There's more space surrounding my head. My jaw relaxes, no longer clenched in anticipation of my mouth filling with alcohol. I know that quitting means sweat-soaked bedsheets, convulsions, confusion, spiders in my wrists, yet I'm relieved for my future self.

Time feels very strange. I've arrived. Again. Here, now, present, in a way I wasn't only seconds ago. The sober self I've been before – as a child, in previous periods of recovery – is no longer a fuzzy memory, lost in the past, but alive again, part of me.

In Thesis XV, Benjamin writes: "The awareness that they are about to make the continuum of history explode is characteristic of the revolutionary classes at the moment of their action."[25] French revolutionaries launched a new calendar after removing King Louis' head. They felt themselves transforming not only a political system but time itself. When fighting began in the uprising of 1830, revolutionaries shot out clocks in public squares across Paris. The time of Restoration was over. Returning from exile to revolutionary Russia, Lenin declared: "There are decades when nothing happens, and weeks when decades happen." Empty time of decades and the explosion of revolution. He wrote that revolutions have their own tempo, which shouldn't be understood as the concluding point of step-by-step progress: "Gradualness explains nothing without leaps. Leaps! Leaps! Leaps!"[26] In 2011, revolutionaries in Tahrir Square blew open space and time in central Cairo. Decades of empty time under the dictator Mubarak were exploded in a matter of weeks. Day after night after day without pause, millions of Egyptians chanted and sang, fed each other, ran schools and clinics, and fought back Mubarak's goons. In Lenin's language, the Arab Spring was a leap.

For *Blackout* author Sarah Hepola, sobriety began with the decision "to quit drinking for one day. And then I tried a month. And then six."[27] Writer and Canadian senator David Adams Richards, who hallucinated dragonflies when he was drunk, went to an AA meeting before committing to sobriety.[28] Not me. My movement from using to sobriety is a leap from the old world to the new. I've let go of whatever I was clinging to or grabbed hold of something I'd lost. I've quit. In Thesis XVI, Benjamin envisions a tipping point in the revolution at which "time stands still and has come to a stop."[29] The old regime is no more.

I don't know why I quit precisely then, not a week (or a minute) earlier or later. It's as though the decision was made beyond my conscious control. Not that a voice speaks to me. I don't surrender to a higher power. When Bill Wilson, the cofounder of AA, quit drinking, he "felt lifted up, as though the great clean wind of a mountain top blew through and through."[30] I know the ecstasy of quitting. What I don't share is Wilson's belief that God made quitting possible, or that "without Him I was lost." Unlike Big Bill, I haven't entrusted my sobriety to "His care and direction."[31]

It's me who quits. (In Michael Löwy's book on Benjamin's philosophy of history, he notes that "there is no Messiah sent from Heaven: we are ourselves the Messiah."[32]) Yet the knowledge that I am quitting appears fully formed. No deliberation, no bargaining for a trial period. When I deliberate and bargain, I live in the blackout. When the awareness that I'm ready to stop dawns on me, I have already stopped. It's a break in being, in time, as abrupt and profound as moving in and out of blackout. But through this gate I move with eyes open.

What will I do with the hours I reclaim? How long can I keep this up? I mark the date: January 15, 2005. October 1, 2013. May 23, 2018. I mark each sober year with a thin, black tattoo. I have nine such tattoos on my elbows and fingers, which is not to say nine sober years in a row, but nine of the past twenty years. The lines are messages to my future self about what needs doing and that I can do it. Confident and disbelieving, I think of Patrick Lane's words: "Daily I feel amazed I am not drinking."[33]

In Benjamin's terms, time in sobriety is filled by the presence of the now. Events and relationships are meaningful in themselves, not defined by the extent to which they enable or restrict drinking. Because my identity doesn't snap in half each night, I am capable of forming a life with narrative continuity. This is essential to what Richard Sennett calls "character [. . .] a sense of sustainable self."[34] Who knows what the day will entail? Except this: it will not end in blackout. For the first time since the start of my last drinking period, life contains the potential of variance. Hope rises in the future.

What about the past, though, all that time lost to drinking? In sobriety, I'm haunted by Raymond Carver's poem "Alcohol." A man hears a song he knows means something to him: "But you don't remember / You honestly don't remember."[35] What do I do with alcoholic memories never made, memories unrecallable and memories I wish I couldn't recall? In her memoir *October Child*, Linda Boström Knausgård asks, "What is the weight of memory? How can it be measured? How do you assign value to memories?"[36] She's mourning her memories destroyed by electroconvulsive therapy. I lost time by my own hand.

Am I obligated to renounce every memory from my blackout years because all of them are marked by the violence of my drinking? At times that seems logical to me, even perversely honourable, although it, too, seems recklessly self-abusive. Even sprinting toward death in the blackout, I was always, like all of us, partially driven by the forces of life-making. Inside the drinking frame, I visited my grandparents, read the newspaper, marched against capitalism. I learned every word to every song by the Beatles, served soup at a homeless shelter, played my guts out in summer softball leagues. I played guitar and sang Christmas carols for seniors at long-term care homes. I became a professor. Fell in love with the woman I want to spend my life with. I laughed a lot.

Much of this is unremarkable, normy stuff. That's exactly my point. There was lightheartedness, empathy, curiosity in the me who drank to blackout. Periodically, I tried to connect, to love, to help others, to revel in the social nature of being human. Even in the face of my alcoholic death drive, I was moved by the compulsion to live.

The trouble is, in sobriety, I've never been able to connect with the life-affirming episodes of my past. Focusing on the "good" things I did in periods of drinking feels like ignoring the violence of the blackout, if not preparing my slide into relapse. So I once helped an old lady cross the street as a hungover twentysomething.

I made an ethical decision here and there; found innocent pleasures between blackouts. Who cares? Whenever I was using, whether drinking that moment or not, my life instincts were subjugated to the death drive. Whatever else I was doing, I was drinking myself to death. (I guzzled Jim Beam from a mickey in the bathrooms of the long-term care homes I carolled in. My sweat on the softball field stank of whiskey.) My joy, care for others, gratitude and honesty were aberrant and unfulfilled – vanquished, really. My life-making efforts were scattered, unachievable episodes of an incoherent, ulti-mately failed drive. Finding pleasure in happy memories within the blackout years requires deception by selectivity.

When Caroline said to me that day in Saskatoon, "You'll do this again, and again, and again. You'll do this your whole life," she was right. In the empty time of active drinking, the blackout consumed everything. My hopes and promises and best attempts were stillborn in darkness. Because that was true then, I assumed it must be true of then always, even if I never drink again. Benjamin suggests it could be otherwise: that in conditions of revolutionary change, I may forge meaningful attachments, a continuous identi-ty, with the life-affirming, life-honouring spirit and occurrences of my past.

In Benjamin's conception of time, which Löwy calls "open history," past episodes of life-making can be salvaged from his-tory's wasteland.[37] The revolution not only begins history anew but reawakens long-lost forces of freedom and justice that were defeated in their time. The authority of the conqueror, smashed by revolutionary forces, no longer determines the stories of what hap-pened before. History is reopened by current struggles for justice, which Benjamin says provide "a revolutionary chance in the fight for the oppressed past."[38] Revolutionaries of today draw inspiration from "enslaved ancestors," and rescue history's victims from obliv-ion.[39] In a future revolution for socialist democracy, the "losing" revolutionaries of the Paris Commune, the communist opposition to Stalin, Patrice Lumumba, Malcolm X, 68ers and the Occupy movement become the ancestors of victory.

That's not to say that the revolution erases the agony of failed struggles. On the contrary, in revolution, histories of suffering and the losses of the oppressed are brought to the centre of collective life, remembered fully and honoured for the first time. The victims of the past are absorbed into presentness, given new life through the fight for justice. In Benjamin's words, Messianic time "comprises the entire history of mankind in an enormous abridgement."[40]

I think of the best of the boy I was before taking my first drink. He memorized quotations from *Bartlett's*, performed as a clown at his sister's birthday party, passed the ball to weaker players in gym class, felt ill at the thought of hurting anyone. In seventh grade, after watching a video by Mothers Against Drunk Driving, he solemnly swore never to drink and drive. He was open to realizing his full human potential.

When I live inside the empty time of the drinking frame, there can be no link, no continuity between the life-destroying alcoholic I am each moment and that life-loving boy. The boy's promise, his dreams, his capacities are impossible. They disappear into blackout. By courting oblivion, I surrender my claim to that boy's legacy. However, in the moment of quitting – the revolution of the life instincts – the best of that boy is redeemed. Past efforts at living well are no longer failures; they are reawakened and become capable of fulfillment. To paraphrase Benjamin, the dawn of sober time comprises the entire history of my existence in an enormous abridgement. Death is not extinguished but integrated with a new whole in which life is possible.

I think of the father I was to Gus in the 142 days between his birth and my last drink, the hours I conceded to the oblivion of alcohol. I lost nights sleeping beside my boy because my blacked-out body threatened his safety. I'd wake in the basement not remembering whether I'd chosen to sleep alone, or my partner forced me from the bedroom.

But my being drunk isn't the whole story of those 142 days. I also pressed Gus's tiny hand to the smooth coldness of windows, the rough dampness of bark, because I wanted him to meet the world. I showed him Halifax Harbour on a cold, sunny morning. I didn't drink at all the week Gus was born. The life instincts lived in these moments, but they drowned in the context of my drinking. Green shoots die in toxic soil. My addiction killed new growth.

When I reflect on my past with the help of Benjamin, I can seize hold of those reborn life-making memories and wrest them from the grip of death drinking.

Redeeming the spirit of living well in that boy, that new father, all my past selves, is no mere matter of looking at things differently and telling myself a new story. Redemption in Benjamin's terms requires substantive, embodied transformation: a radically changed material world, from which new ways of being emerge. In my case, as long as I'm still inside the drinking frame, no amount of remorse, good intentions, generosity, making amends or forgiveness could redeem lost forces of life-making in my past. It is the revolution, the arrival of the Messiah, the moment of quitting – the actual *material* change in my life: a mind and body free from alcohol – that reopens history.

The blackout can't be erased; but from the vantage point of sobriety, its power over the past is not what it was when I was drinking. From where I stand today, blackouts might be the defeated side in a struggle they once dominated completely. If this is true, then the compulsion to live well, which existed at one time only to be vanquished by alcohol dependency, rises again in the past to become the forerunner of liberation. This doesn't cancel the pain of time gone. Harms linger. Grief remains. But whereas I once asked "What do I do with my blackouts?" as a way to say that nothing could be done, I've found in Benjamin the possibility of recovering parts of what's been lost. It seems that the crisis, once permanent, can be transcended.

Theories of crisis typically ignore the impact of decisive events on time gone by. In psychology and sociology, the questions are always: What caused the crisis? And how will it end? Breakdown? Renewal? Transition to a new system? Focus is restricted to the future. Benjamin understood that crises also have the potential to change the past. The past is an ongoing struggle, says Benjamin, neither a linear story of progress nor decline. It is a series of images whose meanings to us are contingent upon the dangers and needs of our moment. Our job, says Benjamin in Thesis VI, is "to seize hold of a memory as it flashes up in a moment of danger" and use the present-packed images of history both to resist the catastrophe of today and prepare for the eruption of unforeseen possibilities through which we can make the world more humane.[41] In her short biography of Benjamin, Esther Leslie interprets his life and work "not only as a register of horror, but also an index of possibilities yet to be realized – and an instrument toward their realization – once we all begin to truly live."[42]

Of course, as Benjamin knew, changes in the future may change the past again. The French revolutionaries of 1789 were thrown back into despotism under Napoleon. Soviet democracy, a flash of hope in 1917, was crushed within a decade by Western aggression and Stalin's rule. In Egypt, the Arab Spring ended with El-Sisi's counterrevolution. I may relapse, fall back into the empty time of alcoholism. And if I do, I will sever the connective tissue forming between the revolution of quitting and moments of living-well earlier in my life. I will no longer be able to identify with the life-making potential of those images in my past because my existence will ensure their destruction. Should the death drive of blackout become dominant again, the meaning of those moments in which I've embraced life before will again become fragmented, isolated, unrealizable. It will take another explosion of history to reunite them into a coherent identity and redeem the fullness of life against death.

I write on a laptop in my home office. It's unseasonably warm for May, so I'm wearing shorts and a T-shirt. Through the window to my left, I see ivy-covered wood fences, lilacs in bloom, the green shingled roof of my neighbour's toolshed. Grackles peck at mulch beneath a swing set.

In my office, books fill shelves along two walls. On the other walls hang framed bird prints, a plaque from a Kandinsky exhibition, a small reproduction on wood of the Paul Klee painting that inspired Walter Benjamin's angel of history. My orange cat sleeps on a brown leather chair. Now and then I catch a whiff of cinnamon from this morning's French toast.

How dare I compare my personal struggles to those of Benjamin's suffering masses and revolutionaries? In aid of my therapeutic needs, I've cast the best parts of me sabotaged by drinking as Benjamin's peasants, workers, slaves and refugees: the oppressed held down and slaughtered in the millions by the conquerors of history. My blackouts are the empty time of class society. The delusional philosophy of history as progress is my own small conception of maturing within the drinking frame. My compulsion to drink is the butchering ruling classes. The life instincts in me are the ineffable forces of bravery, sacrifice, justice and freedom on a mass scale. The Messiah comes when I quit drinking.

Is this narcissism? Benjamin is writing of social classes that are, by definition, at war with each other. Anti-fascists and Nazis, working class and capitalists – oppressed and oppressor. How could both exist in me, an individual?

The sociological objection doesn't bother me. Benjamin was a bold dialectical thinker. On the one hand, he viewed the whole of society in terms of discrete parts, more or less in conflict with other parts; and on the other hand, he envisioned the parts in terms of their relations within a social whole. His life's work dwells on the particulars, contradictions and conflict among social groups in flux, yet it also dreams of integration through socialist revolution. There is room in Benjamin's dialectical sociology to experiment with ideas about how the parts and whole relate – in society, in individuals, among the two. But rather than go too far with a defence on the terrain of social science, I'll point out that my use of Benjamin's model turns on analogies and metaphors, not literal interpretations, not clones. The power of metaphor lies as much in difference as in similarity. The question is whether combining different materials creatively lights new sparks by which to see better.

I'll admit it: the Lord does not come when I quit drinking. Don't forget, though: Benjamin's Messiah was not actually the Lord. Benjamin's work was experimental, iconoclastic. He sought synthesis where others saw no connections. True to Benjamin's belief in the educational power of montage, aphorism and contradiction, his "Theses on the Philosophy of History" draws on imagery from folktales, biblical revelation, modern painting, ancient Rome, steam trains, analyses of light, fashion and bordellos. He wrote essays about theatre, children's literature, smoking hash, urban design, dreams and fascism. He was among the first to host a radio show, a post he used to test methods of Marxist popular education. His "A Berlin Chronicle" is at once a record of Benjamin's early years, a phenomenological map of the *fin de siècle* city and a meditation on the nature of memory. Little would be more alien to Benjamin's legacy than policing intellectual boundaries.

I say we who call ourselves radicals ought to assume the role of heirs to the revolutionary spirit in all aspects of our lives. We do so openly in the political sphere: sticking rose emojis to our Twitter handles, quoting Marx on May Day, celebrating the Stonewall riot, raising solidarity fists in the street. Like good Benjaminians, we brush history against the grain, seizing hold of images of resistance as they flash up from history in our moment of danger. We do so to show respect, forge bonds and exercise collective learning. To conclude that revolutionary histories and theory cannot be material for working through personal struggles, my alcoholism, in this case, because the stakes and scale of analysis are so different from the political sphere, strikes me as anti-revolutionary: counter to the openness, generosity and warmth that drives the tradition of democracy-from-below. Certainly, Benjamin's writing life was devoted to locating and amplifying life-making forces amid the war with the death drive everywhere.

My bigger concern with reading recovery through Benjamin's philosophy of crisis is the possibility that I'm drawn to his model not because it holds unique insights on quitting but because it enables my alcoholic tendencies. Leslie Jamison is similarly concerned about her attachment to all-or-nothing solutions. Toward the end of *The Recovering*, Jamison writes about what a drag sobriety can be. You get sober, yet you still have so many problems – from money trouble to heartache to everyday ennui. Now you have new problems: like what to do every night, and the fact that your drunk friends are boring. You haven't written the novel you said you would write. Hobbies are dumb without a buzz.

So Jamison fantasizes about relapsing, not just to experience the thrill of drinking again, but to "find another bottom, an explosion whose rubble I could emerge from, ash-dusted and glittering with shards of glass. The fantasy of crisis – or explosion – was an alternative to the hard, ordinary work of living through uncertainty; and sobriety had given me fewer explosions to recover from."[43] (It's noteworthy that Jamison's "explosion" is the depth of the crisis, whereas, for Benjamin, it's the moment of breaking free. There is comfort and fear in hitting rock bottom, just as there is in liberation.) The writer Lorna Crozier says her life-long relationship with Patrick Lane – "a stormy journey of lovemaking, fights, revelries and poems" – was born in a shared passion for drinking: "We thirsted for intensity; we savoured crises."[44]

Do I fetishize Benjamin's theory of history as crisis and explosion because I'm still attached to the highs and lows of blackout drinking? Maybe. Apart from falling in love and having children, I've never experienced more profound transformations than coming in and out of blackout, in and out of the drinking frame. It could be that I cling to the idea of quitting as the coming of the Messiah because I reserve the right for my next relapse (please, Lord, don't let it come) to be a spiral into hell: destined for the bottom, impossible to slow. I wonder why I identify with imagery in Benjamin's essay more easily, more viscerally than with portrayals of recovery in the stacks of quit lit on my bookshelves. It's not as though in sobriety I'm freer from the crises of history than are Jamison, Hepola or any other recovering addict. The satisfaction available in society is constrained by social structures of injustice. "Real satisfaction," writes Beaupre, in his article on recovery and Benjaminian storytelling, "would amount to a change in the conditions whereby one suffers."[45]

I've leapt from the flames and landed on my feet, but where have I landed exactly? In the crisis of capitalism, the crisis of ecology. In the crisis of modernity, as diagnosed by Benjamin's friend Hannah Arendt. Arendt said that the Western tradition, as a project of social cohesion, is dead. Capitalism, bureaucracy and war have eroded the shared cultural assumptions necessary to hold together community across vast lands and large populations. In the ruins of Western religious and political authority, the masses grope for meaning in a state of "worldlessness."[46]

Arendt believed that this crisis could be resolved through radical world-rebuilding rooted in participatory democracy.[47] She warned, however, that until new worlds of freedom and civic action are established, the crisis of modernity leaves societies vulnerable to totalitarian movements. Fascism thrives in conditions of mass disorientation.

Arendt's crisis of modernity hasn't gone away, and the far right is ascendant. When I envision myself stepping from one crisis (of alcoholism) to another (of modernity, climate change, capitalism, whatever), I can be overwhelmed by a sense of futility. There is no escaping crisis. Why bother trying?

Yet, breaking free of the blackout has been crucial to confronting the social crises that Benjamin and Arendt are talking about. Gaining basic functionality is part of it. I'm no longer so fuzzy with drink or withdrawal that I skip the meeting, lose the thread, withdraw into myself. More importantly, though, through the revolution of quitting, I've recaptured my sense that what matters is "living on" in accordance with a moral vision, to quote Hägglund's *This Life*.[48] My "secular faith" has been restored. One response to capitalist alienation is to disappear into blackout. A different response, also personal and political, is to leap from blackout to social life. Social life, too, is crisis-ridden, alienated: marked by inequities, complexities and uncertainties. But in communion with others is the only place any of us can try to balance our individual angel-of-self with larger angel-of-history concerns. What to do in the future with the wreckage of my past is my plight, but isn't that the plight of all of us?

For Hägglund, the heart of secular faith is "the sense of the ultimate fragility of everything we care about," and the effort to protect and prolong the life of these things – from loved ones to political projects – in spite of knowing (Hägglund argues *because we know*) that they are finite; they will pass away.[49] Once we've accepted that nothing lasts forever, the essential question for Hägglund is: How should we spend our time together? This afternoon, this decade; you and me, the whole world; as we organize a picnic, as we organize the economy – over a lifetime, "in *this life* as an end in itself."[50]

I cannot answer Hägglund's question, can't even contemplate it, when I live inside the drinking frame. The question is negated as soon as it's uttered because of my attachment to the empty time of blackout. Blackout courts death, is death; and death is eternal – empty time. By breaking free of blackout, I have a very different stake in my own life, which is also always a life-in-society. I begin to see patterns in crises that manifest in external control of our life-making powers and the experience of empty time: in alcoholism, in capitalism (in struggling with addiction under capitalism). How I conceive of, and deal with, one (personal) crisis overlaps with my conception of, and ability to grapple with, other (social) crises. Personal crises unfold within specific social conditions. Society will be transformed (or not) by people managing personal crises.

I'd be disappointed to be understood as saying little more than "get your own shit together before you try to change the world." We all have shit, and not all of it will ever be taken care of. I applaud all who struggle for social justice, no matter how torturous their personal demons. I'm not even writing to urge other addicts to quit using. I assume that people have a pretty good sense of what they want, what they're capable of and what's best for them, in the context of their unique burdens and supports. What I'm saying is that unexpected insights may be found in reflecting on crises at different levels of experience in relation to each another. My recovery has been bolstered by reconceiving of my blackouts as Benjamin's empty time. I've begun thinking of the periods between social revolutions as the blackouts of history. I watch for the sparks, feed flashes in darkness, that might be the next explosion, dare I say the final break from the eternal recurrence of the same?

This time I'm quitting for real: the great alcoholic cliché. We know Benjamin's angel of history has seen the final revolution erupt before, the actual dawn of systemic change, only to see catastrophe return. Yet I cannot accept that my life is destined to cycle through relapse and recovery. To accept as much would already be to put the bottle back in my hand. I imagine myself as the angel – not the angel of history; call it my patron saint of drinking – closing my wings against storms from the past, at last turning toward the future. Celebratory sobriety tattoos cover his elbows and fingers. Images of now-time from the past piecing together what's to come.

NOTES

1   Leslie Jamison, *The Recovering: Intoxication and Its Aftermath* (Little, Brown, 2018), 26.

2   Donald W. Goodwin, "Alcohol Amnesia," *Addiction* 90, no. 3 (1995): 316.

3   Donald W. Goodwin, J. Bruce Crane and Samuel B. Guze, "Phenomenological Aspects of the Alcoholic 'Blackout,'" *British Journal of Psychiatry* 115, no. 526 (1969): 1034, italics in original.

4   Marlon P. Mundt and Larissa I. Zakletskaia, "Prevention for College Students Who Suffer Alcohol-Induced Blackouts Could Deter High-Cost Emergency Department Visits," *Health Affairs* 31, no. 4 (2012): 863–70.

5   Sarah Hepola, "Prelude: The City of Light," in *Blackout: Remembering the Things I Drank to Forget* (Grand Central Publishing, 2015), Kobo.

6   Hepola, *Blackout*, italics in original.

7   Karl Marx, *Economic and Philosophic Manuscripts of 1844*, trans. Martin Milligan (1844; Prometheus Books, 1988), 76–78, italics in original.

8   Martin Hägglund, introduction to *This Life: Secular Faith and Spiritual Freedom* (Pantheon Books, 2019), Kobo.

9   Patrick Lane, *There Is a Season* (Emblem, 2009), 43.

10 Alcoholics Anonymous, *Alcoholics Anonymous: The Story of How Many Thousands of Men and Women Have Recovered from Alcoholism*, 4th ed. (A.A. World Services, 2001), 59.

11 Saidiya Hartman, *Lose Your Mother: A Journey Along the Atlantic Slave Route* (Farrar, Straus and Giroux, 2007), 16.

12 Joel C. Beaupre, "Storytelling: Walter Benjamin and Recovery from Alcoholism," *Contemporary Social Science* 8, no. 1 (2013): 33.

13 Beaupre, "Storytelling."

14 Walter Benjamin, "Theses on the Philosophy of History," in *Illuminations: Essays and Reflections*, trans. Harry Zohn, ed. Hannah Arendt (Mariner Books, 2019), 200.

15 Quoted in Esther Leslie, "Time for an Unnatural Death," chap. 7 in *Walter Benjamin* (Reaktion Books, 2007), Kobo.

16 Victor Serge, *Midnight in the Century*, trans. Richard Greeman (1939; New York Review of Books Classics, 2014).

17 Benjamin, "Theses," 202.

18 Benjamin, "Theses," 201.

19 Benjamin, "Theses," 200.

20 Benjamin, "Theses," 205.

21 Benjamin, "Theses," 206.

22 Jamison, *Recovering*, 425, italics in original.

23 Benjamin, "Theses," 206.

24 Benjamin, "Theses," 205.

25 Benjamin, "Theses," 206.

26 Quoted in Daniel Bensaïd, "'Leaps, Leaps, Leaps': Lenin and Politics," *International Socialism*, no. 96 (July 2002), Marxists' Internet Archive, last updated June 19, 2012, https://www.marxists.org/archive/bensaid/2002/07/leaps.htm.

27 Sarah Hepola, "Bingeing on Not Drinking," *Atlantic*, January 8, 2011, https://www.theatlantic.com/daily-dish/archive/2011/01/bingeing-on-not-drinking/177548/.

28 David Adams Richards, "Drinking," in *Addicted: Notes from the Belly of the Beast*, ed. Lorna Crozier and Patrick Lane (Greystone Books, 2006), Kobo.

29 Benjamin, "Theses," 207.

30 Alcoholics Anonymous, *Alcoholics Anonymous*, 14.

31 Alcoholics Anonymous, *Alcoholics Anonymous*, 13.

32 Michael Löwy, "A Reading of Walter Benjamin's 'Theses "On the Concept of History,"'" chap. 1 in *Fire Alarm: Reading Walter Benjamin's "On the Concept of History,"* trans. Chris Turner (Verso, 2016), Kobo.

33 Lane, *There Is a Season*, 47.

34 Richard Sennett, *The Corrosion of Character: The Personal Consequences of Work in the New Capitalism* (W.W. Norton, 1998), 27.

35 Raymond Carver, "Alcohol," in *Fires: Essays, Poems, Stories* (Vintage Books, 1984), 47.

36 Linda Boström Knausgård, *October Child*, trans. Saskia Vogel (World Edition, 2020), Kobo.

37 Löwy, "The Opening-up of History," chap. 2 in *Fire Alarm*, Kobo.

38 Benjamin, "Theses," 207.

39 Benjamin, "Theses," 204.

40 Benjamin, "Theses," 208.

41 Benjamin, "Theses," 198.

42 Leslie, "Benjamin's Finale: Excavating and Remembering," in *Walter Benjamin*, Kobo.

43 Jamison, *Recovering*, 426.

44 Lorna Crozier, "Breathing Under Ice," in *Addicted: Notes from the Belly of the Beast*, ed. Lorna Crozier and Patrick Lane (Greystone Books, 2006), Kobo.

45 Beaupre, "Storytelling," 34.

46 Hannah Arendt, "Remarks on the Crisis Character of Modern Society," *Christianity and Crisis* 26, no. 9 (1966): 113.

47 Hannah Arendt, *Between Past and Future*, rev. ed. (Penguin, 2006).

48 Hägglund, "Love," chap. 2 in *This Life*.

49 Hägglund, introduction to *This Life*.

50 Hägglund, *This Life*, italics in original.

ACKNOWLEDGEMENTS

I was able to write this book because of the encouragement and guidance of Wolsak and Wynn's publisher, Noelle Allen. From the first time I described to Noelle my interest in writing about our times of crisis and our collective interest in talking about crises, she understood the project completely and saw possibilities for pursuing it that hadn't occurred to me. It's a rare and wonderful thing to work with an editor who truly *gets* what you're trying to do, even when what you're trying to do remains imperfectly conceived and only ever quasi-articulated. Whether Noelle was reassuring me about the quality of a draft or inviting me to consider dropping a piece that I'd previously held close, at every phase of the writing process, she was engaged in the details of my work and honoured the project's overall vision. I will forever be grateful for Noelle's superlative skills in the literary arts and the generosity with which she shares them.

I am grateful, too, to Ashley Hisson and Jen Rawlinson of Wolsak and Wynn, and Hollay Ghadery of River Street Writing for making the book beautiful and helping share it with the world. Thank you for the scrupulous copy edit, Megan Beadle. Thanks also to the publications that printed earlier versions of a handful of the essays appearing in this book, namely: *Best Canadian Essays 2025* (ed. Emily Urquhart, Biblioasis), *Canadian Notes & Queries*, *Canadian Journal of Communication*, *Hamilton Review of Books*, *Socialist Studies*, *TOPIA: Canadian Journal of Cultural Studies* and the *Montreal Review of Books*.

It would be unconventional, not to mention impossible, to name every intellectual influence to which this book is indebted, in all their varying states of solidity. (Thank you, Marxism? Thank you, rivers near my home? Thank you, printing press and farm workers and dream world?) Yet because I think of it on nearly a daily basis, I want to acknowledge the influence on my thinking of Italian author Elena Ferrante's way of talking about the social character of every book, every work of art, every idea. In Ferrante's words, "There is no work of literature that is not the fruit of tradition, of many skills, of a sort of collective intelligence. We wrongfully diminish this collective intelligence when we insist on there being a single protagonist behind every work of art." With that thanks to the soil, the philosophers, the caregivers, the clouds, the birds and the readers now leading the way, I can thank the indubitably solid many friends who generously offered feedback on parts of the manuscript while it was still a work in progress. Thank you, Kate Cairns, Rick Cairns, Jessica Altenburg, Sue Ferguson, Alan Sears, Layne Beckner Grime (thanks for the author photo, too, Layne!), Alison Fishburn, Dana Hansen, Gary Barwin, Todd Gordon, David Camfield and Dan Snaith. Dan provided probing commentary on the entire manuscript, some essays he critiqued more than once, while composing another brilliant album of his own and touring mega music venues around the world. It's a pleasure to pay special tribute to the intelligence, loyalty, fervour and playfulness of Dan's friendship.

Nothing I do could I ever have done without the feeling of being loved and actively supported by my family. Dad and Beth, Mom and Wen, Grapegrammamu and Oma and Opa and Leslie, Simon and Teresa, Kate and Corey, and all other Altenburg-Jacobs-Duggan-Cairnses out there, thank you for all that you give and give and give to me.

My greatest thanks of all is offered to the three people who have influenced the book most profoundly, and for whom the book is written. Jessica, Gus and Winnie, you are my window on the world, illuminating all I see, while filling with light my heart, my brains, my bones, guts, and laughter, even in the dark times. I doubt anyone but me will ever fully understand just what I mean when I say this book is not only possible because of you (the ideas you've given me, the time you've given me, the hope you spark in me, the fear I feel because of my love for you), but that really, this book is about you, about us, about the early years of our lives together, even when it doesn't read that way at all. A writer only ever writes from a specific place and time, and you are the place and time from which I write about the world. What a wonderful place and time to be in. Jess, you are everything to my writing life, and to all my life not written. May you and I be tangled up together from crisis to crisis to crisis, and in every boring bit between.

James Cairns lives with his family in Paris, Ontario, on territory that the Haldimand Treaty of 1784 recognizes as belonging to the Six Nations of the Grand River in perpetuity. He is a professor in the Department of Indigenous Studies, Law and Social Justice at Wilfrid Laurier University, where his courses and research focus on political theory and social movements. James is a staff writer at the *Hamilton Review of Books*, and the community relations director for the Paris-based Riverside Reading Series. James has published three books with the University of Toronto Press, most recently, *The Myth of the Age of Entitlement: Millennials, Austerity, and Hope* (2017), as well as numerous essays in periodicals such as *Canadian Notes & Queries*, the *Montreal Review of Books*, *Briarpatch*, *TOPIA*, *Rethinking Marxism* and the *Journal of Canadian Studies*. James' essay "My Struggle and *My Struggle*," originally published in *CNQ*, appeared in Biblioasis's *Best Canadian Essays 2025* anthology.